Lecture Notes in Mathematics

A collection of informal reports and seminars
Edited by A. Dold, Heidelberg and B. Eckmann, Zürich

Series: Institute of Mathematical Sciences, Madras
Adviser: K. R. Unni

232

Tord H. Ganelius

University of Göteborg, Göteborg/Sweden

Tauberian Remainder Theorems

Springer-Verlag
Berlin · Heidelberg · New York 1971

AMS Subject Classifications (1970): 40 E 05, 44 A 35, 41 A 25, 42 A 92

ISBN 3-540-05657-2 Springer Verlag Berlin · Heidelberg · New York
ISBN 0-387-05657-2 Springer Verlag New York · Heidelberg · Berlin

Offsetdruck: Julius Beltz, Hemsbach/Bergstr.

PREFACE

The main part of these notes consists of the material in a series of lectures I gave at the Institute of mathematical sciences "Matscience" in Madras in January 1969 and which has been available in a mimeographed report from that institute. In this edition I have added some topics I have given in lectures at other places. Many theorems represent unpublished work of mine and I have also included some references to work of my students.

The main theme is the application of Wiener's general method to different kinds of tauberian problems, in particular to remainder problems. As will be seen the proofs of general theorems are often simpler than special proofs for special cases.

The first application of Wiener's method to remainders in tauberian theorems was given by Beurling in 1938. His work has been continued by Lyttkens. Among other things she has proven theorems where conditions on the Fourier transform of the kernel are required only in the upper half-plane. Such theorems are certainly important e.g. in number theory. I have chosen to formulate my theorems in a less general way in this respect, but it is easy to see that the method can be used in many cases also under the weaker assumptions.

For help in the preparation of the notes for publication thanks are due to the staff of Matscience but also to J.E.Andersson, A.Ganelius, T.O.Ganelius and K.Lidhag.

August 1971 Tord Ganelius

C O N T E N T S

CHAPTER 1

DIFFERENT APPROACHES TO TAUBERIAN THEOREMS

1.1 Introduction

Let us start with the theorem of Abel: If $(a_n)_0^\infty$ is a real sequence, such that

(1) $$\sum_0^\infty a_n = 0,$$

then

(2) $$\lim_{x\uparrow 1} \sum_0^\infty a_n x^n = 0 .$$

There is a converse problem: To find conditions on $(a_n)_0^\infty$ so that (2) implies (1). Results in this direction are called tauberian theorems after A.Tauber, who in 1897 proved that

(3) $$na_n = o(1)$$

is a 'tauberian condition' of this type. That some condition is needed is seen by the simple example

$$\sum_0^\infty a_n x^n = \tfrac{1}{2}(1+x)^{-1}(1-x),$$

where $a_0 = \frac{1}{2}$, $a_n = (-1)^n$ if $n > 0$.

The theory was developed by Hardy and Littlewood, one of the most famous results being the following

THEOREM (Hardy and Littlewood 1913):

If (2) holds and

(4) $$na_n \leq 1,$$

then (1) follows.

Hardy and Littlewood used many ingenious devices and their proof of the result just mentioned is quite complicated. In 1930 J.Karamata published a remarkably simple proof of one of the fundamental lemmas, thus considerably simplyfying the proof of the theorem.

A few years later N.Wiener's large paper 'General Tauberian theorems' appeared in Annals of Mathematics. In this work the original problem of Tauber and a number of other problems of similar character, which had been handled by different methods, were shown to be particular cases of a general theorem,

WIENER'S TAUBERIAN THEOREM:

If $K \in L(-\infty,\infty)$, $\phi \in L^\infty(-\infty,\infty)$ and

$$\hat{K}(t) = \int \exp(-2\pi ixt)K(t)dt \neq 0$$

<u>for all real t, then</u>

(5) $$K * \phi(x) = \int K(x-y)\phi(y)dy = o(1), x \to \infty,$$

<u>implies</u>

(6) $$H * \phi(x) = o(1), x \to \infty,$$

<u>for every</u> $H \in L(-\infty,\infty)$. (Unspecified integrations refer to the whole space.)

The key stone in Wiener's proof is the

<u>LEMMA:</u>

<u>If</u> $\hat{K}(t) \neq 0$ <u>for all real</u> t, <u>then the function</u> m <u>defined by</u> $m(t)\hat{K}(t) = 1$ <u>is locally a Fourier transform.</u>

That means that to every real number τ there is a neighborhood O_τ and a function $M \in L(-\infty,\infty)$, such that $\hat{M}(t) = m(t)$ for $t \in O_\tau$.

Gelfand's work on normed rings (1939) revealed that this lemma can be proved in a comparatively simple way in a much more general setting. In this general form the tauberian theorem is an important result in the generalized harmonic analysis on locally compact groups, thus being one of the first instances of the succesful interplay between more modern algebraic geometric methods and classical analysis.

Tauberian remainder theorems certainly belong to the field of classical hard analysis. We shall however find that certain result in weak analysis are very useful. The connection with approximation theory will be evident at most points.

To give a background and for purposes of reference, I shall give the main parts of three different proofs of Hardy-Littlewood's theorem, one by approximation, one by Wiener's method and one by Schwartz's theory of distributions. All can be modified for applications to problems on remainders.

This might be the natural place to state a typical remainder problem. Let us first consider Hardy-Littlewood's theorem.

Let us assume that the rate of decrease to zero of $\sum_0^\infty a_n x^n$ is known as $x \uparrow 1$ for instance, that

(7) $$\sum_0^\infty a_n x^n = O((1-x)^\delta), \ \delta > 0,$$

and also that (4) holds. Is it then possible to improve the conclusion (1) ? The answer is affirmative and it was first proved by Freud in 1951 that these assumptions imply

(8) $$\sum_{n<x} a_n = O(1/\log x), \ x \to \infty.$$

The crucial example showing that the result cannot be improved was given by

J.Korevaar (1951) who also published a series of results in the field in 1953-54.

Our central theme is the

GENERAL TAUBERIAN REMAINDER PROBLEM:

To characterize pairs of sub-sets $(\mathcal{L}, \mathcal{M})$ of $L(R^d)$, $d \geq 1$, such that, if $K \in \mathcal{L}$ and

$$K * \phi(x) = O(r(x)), \quad x \to \infty,$$

then

$$H * \phi(x) = O(s(x)), \quad x \to \infty,$$

for all $H \in \mathcal{M}$, where r and s are functions tending to 0 as x tends to ∞.

More definite statements will follow in the third chapter. My method is a modification of Wiener's and by simple methods of functional analysis it can in many cases be shown that the results obtained are precise. Hence we avoid the troublesome construction of counter-examples. (I am indebted to L.Hörmander for drawing my attention to this fact.)

1.2 Approximation methods

In 1952 H.Wielandt published a very nice modification of Karamata's proof, which gives a short direct proof of Hardy-Littlewood's theorem. It is a typical proof by approximation theory and it can easily be adopted to give the best possible remainder results for power series and Laplace transforms.

Wielandt considers the space $\mathcal{F}$ consisting of those functions f for which $\sum a_n f(x^n)$ converges and

$$(\zeta) \qquad \sum_0^\infty a_n f(x^n) \to 0, \quad x \uparrow 1.$$

According to the assumption (1.1.2) of Hardy-Littlewood's theorem every power of x belongs to $\mathcal{F}$ and since $\mathcal{F}$ is a linear space, every polynomial without constant term belongs to $\mathcal{F}$. What we want to prove is, that the function g defined by

$$(2) \qquad g(x) = \begin{cases} 0, & 0 \leq x < \frac{1}{2}, \\ 1, & \frac{1}{2} \leq x \leq 1, \end{cases}$$

belongs to $\mathcal{F}$. Since $\sum_0^\infty a_n g(x^n) = \sum_0^{N(x)} a_n$, where $N = \left[\frac{\log 2}{\log(1/x)}\right]$, Hardy-Littlewood's theorem follows if $g \in \mathcal{F}$.

We shall prove that $g \in \mathcal{F}$ by showing that every function h that can be approximated by polynomials p_1 and p_2 without constant terms and with $p_1(1) = p_2(1)$ in such a way that

$$(3) \qquad p_1(x) \leq h(x) \leq p_2(x), \quad \int_0^1 q(x)dx < \varepsilon \quad \text{where } x(1-x)q(x) = p_2(x) - p_1(x)$$

belongs to $\mathcal{P}$.

That g can be approximated in this way follows by aid of Weierstrass theorem; Consider $G(x) = x^{-1}(1-x)^{-1}(g(x)-x)$ for $0 \leq x \leq 1$. This function has only one point of discontinuity, $x = \frac{1}{2}$. It can evidently be approximated from above and below by continuos functions s_1 and s_2 in such a way that

$$s_1(x) \leq G(x) \leq s_2(x) \quad \text{if} \quad 0 \leq x \leq 1 ,$$

$$|s_2(x) - s_1 \; x)| \leq \varepsilon \quad \text{if} \quad |x- \tfrac{1}{2}| > \varepsilon ,$$

$$|s_2(x) - s_1(x)| \leq 4 \quad \text{if} \quad |x- \tfrac{1}{2}| \leq \varepsilon .$$

According to Weierstrass we can find polynomials t_1 and t_2 such that $|t_1(x) - s_1(x)| < \varepsilon$, $|t_2(x) - s_2(x)| < \varepsilon$. If we define r_1 and r_2 by $r_1(x) = t_1(x) - \varepsilon$, $r_2(x) = t_2(x) + \varepsilon$,

then

$$r_1(x) \leq s_1(x) \leq G(x) \leq s_2(x) \leq r_2(x)$$

and

$$\int_0^1 (r_2(x) - r_1(x))dx \leq 5\varepsilon(1-2\varepsilon) + (4+4\varepsilon)2\varepsilon \leq 13\varepsilon .$$

It remains to prove that $h \in \mathcal{A}$ if (3) holds. We consider

$$\sum_0^\infty a_n h(x^n) - \sum_0^\infty a_n p_1(x^n) \leq \sum_0^\infty a_n(p_2(x^n) - p_1(x^n))$$

$$\leq \sum_0^\infty \frac{x^n(1-x^n)}{n} q(x^n)$$

$$\leq \sum_0^\infty (1-x)x^n q(x^n).$$

Putting $q(x) = \sum_0^N b_m x^m$ we get

$$(1-x)\sum_0^N x^n q(x^n) = \sum_0^N b_m \frac{1-x}{1-x^{m+1}} \to \sum_0^N \frac{b_m}{m+1} , \quad x \uparrow 1.$$

But $\sum_0^N \frac{b_m}{m+1} = \int_0^1 q(x)dx < \varepsilon$ and hence

$$\overline{\lim_{x\uparrow 1}} \sum_0^\infty a_n h(x^n) < \varepsilon .$$

Starting instead with $\sum_0^\infty a_n p_2(x^n) - \sum_0^\infty a_n h(x^n)$ we obtain

$$\underline{\lim}_{x \uparrow 1} \sum_0^\infty a_n h(x^n) > -\varepsilon .$$

Hence $h \in \mathcal{A}$ and the proof of the theorem is finished.

It is seen from the proof that if we want to obtain a remainder theorem in this way, we must have more precise information about the degree of approximation in its dependence on n. Such quantitative theorems on one-sided approximation have been given by several authors (cf. Freud 1954, Korevaar 1951, Freud-Ganelius 1956) and the field has been developed further in recent years (Bojanic-DeVore 1966).

To show how little is in fact needed to get the precise remainder theorem, I give the following approximation result which can be found in the lit erature quoted above.

APPROXIMATION THEOREM:

Given a function h of bounded variation, then there is a number b only depending on h such that to every ρ there are polynomials p_1, p_2 and q of degree ρ and satisfying

$$p_1(x) \leq h(x) \leq p_2(x), \quad q(x) = x^{-1}(1-x)^{-1}(p_2(x) - p_1(x)), \int_0^1 q(x)dx \leq b/\rho .$$

Moreover there is a number a depending on h such that

$$\sum_{m=0}^{\rho} |c_m^{(1)}| + \sum_{m=0}^{\rho} |c_m^{(2)}| \leq \exp(a\rho),$$

if $(c_m^{(i)})_{m=0}^{\rho}$ are the coefficients of $p_i(x)$.

By aid of this result we can easily prove the remainder theorem mentioned above.

Putting $F(x) = \sum_0^\infty a_n x^n$ we assume that

$$|F(x)| \leq C(1-x)^\delta, \quad \delta > 0. \tag{4}$$

Let p_1 and p_2 be polynomials approximating the function g given in (2) in the sense of our approximation theorem, and let q be the polynomial defined in that theorem.

Then if $N(x) = \left[\frac{\log 2}{\log (1/x)}\right]$ we have

$$\sum_0^{N(x)} a_n = \sum_0^\infty a_n g(x^n) = \sum_0^\infty a_n p_1(x^n) + \sum_0^\infty a_n(g(x^n) - p_1(x^n)) =$$

$$= \sum_{m=1}^{\rho} c_m^{(1)} F(x^m) + \sum_0^{\infty} a_n(g(x^n) - p_1(x^n))$$

$$\leq C(1-x^{\rho})^{\delta} \sum_1^{\rho} |c_m^{(1)}| + \sum_0^{\infty} \frac{1}{n}(p_2(x^n) - p_1(x^n))$$

$$\leq C\rho^{\delta}(1-x)^{\delta}\exp(a\rho) + \int_0^1 q(x)dx$$

as before.

Hence $\sum_0^N a_n \leq \rho^{-1}\{C \exp((1+\delta) \log\rho + a\rho + \delta\log(1-x)) + b\}$

Choosing $\rho = \left[\frac{\delta}{2a}\log\frac{1}{1-x}\right]$ we find that

$$\sum_0^N a_n \leq O((\log\frac{1}{1-x})^{-1}) = O(1/\log N),$$

since $1-x = O(N^{-1})$ by $N = \left[\frac{\log 2}{\log(1/x)}\right]$.

Proceeding similarly with p_2 we get the opposite inequality. Hence we have shown that the remainder result given in (1.1.8) follows from the approximation theorem quoted.

The method applied is essentially the method used by Freud and Korevaar. The most comprehensive statement of the results for Laplace transforms obtained in this way is given in Ganelius 1956 b.

The full proof of my results in that paper has not been published since I discovered that more general results can be obtained by use of Wiener's method (Ganelius 1958, 1962, 1964, Frennemo 1965) and we next turn to that subject. (That my result on the Laplace transform mentioned above, and hence also those of Freud and Korevaar on that problem, can be derived from general theorems in the Wiener sense has been shown by Frennemo 1966, 1967 who also treats the problem in d dimensions).

1.3 Wiener's method

Our next task is to show how Hardy-Littlewood's theorem (and the corresponding remainder theorem) can be transformed to the form treated by Wiener.

In $F(x) = \sum_0^{\infty} a_n x^n$ we put $x = e^{-t}$ and introduce the function $\alpha(\lambda) = \sum_{n<\lambda} a_n$. Thus we obtain

$$F(e^{-t}) = \sum a_n e^{-nt} = \int_0^{\infty} e^{-\lambda t} d\alpha(\lambda) = t\int_0^{\infty} \alpha(\lambda)e^{-\lambda t}d\lambda .$$

We next introduce $t = e^{-u}$ and $\lambda = e^{v}$ and get

$$F(\exp(-\exp(-u))) = \int_{-\infty}^{\infty} \exp(-\exp(u-v) - (u-v))\,\alpha(e^{v})\,dv .$$

If we put $K(x) = \exp(-x-\exp(-x))$, $\phi(v) = \alpha(e^{v})$ and $\psi(x) = F(\exp(-\exp(-x)))$ we find that

(1) $\quad \psi(x) = K * \phi(x) = \int K(x-y)\phi(y)\,dy = o(1), \quad x \to \infty .$

That $\psi(x) = o(1)$ follows since $\psi(x) = F(\exp(-\exp(-x)))$, and $\lim_{x \uparrow 1} F(x) = 0$. We want to prove that it follows that $\phi(x) \to 0$ as $x \to \infty$ under an appropriate tauberian condition. One such condition is: to every $\delta > 0$ there is an $x_0(\delta)$ such that

$$\phi(y) - \phi(x) \leq K\delta,$$

if $x_0(\delta) \leq x \leq y \leq x + \delta$, and with a numerical K.
That this condition is fulfilled if the Tauberian condition (1.1.4) holds, may be seen in the following way.

Since $\sum_{k=m}^{n-1} (1/k) \leq \log (n/m)$ it follows from (1.1.4) and our definition of α that

$$\alpha(y) - \alpha(x) \leq (1/x) + \log (y/x) \quad \text{for } x \leq y.$$

Hence

$$\phi(y) - \phi(x) \leq e^{-x} + (y-x),$$

if $x \leq y$, and it follows that

(2) $\quad \phi(y) - \phi(x) \leq 2\delta \quad$ for $x_0(\delta) \leq x \leq y \leq x + \delta$.

To be able to apply Wiener's theorem, we have to check that $K \in L(-\infty, \infty)$ and that $\hat{K}(t) \neq 0$ for all real t. That is easily done, since a simple computation shows that

$$\hat{K}(t) = \Gamma(1+2\pi it),$$

and Γ^{-1} is an entire function.

We also have to prove that ϕ is bounded, i.e. that the conditions $na_n \leq 1$ and $\sum_0^{\infty} a_n x^n = o(1)$, $x \uparrow 1$, imply that $\sum_0^{N} a_n = O(1)$. This can be proved quite elementary and proofs are found in the classical books on the subject. Instead of writing down another proof I refer to the lines on page 4, where we obtained much more in the proof of Wielandt's theorem. Some methods to handle this problem of boundedness have been given by Vijayaraghavan 1926 (cf. Hardy 1949, p.305 ff). Hardy-Littlewood's theorem now follows from Wiener's in the following way. We recall condition (2) and in the conclusion of Wiener's theorem we take H to be δ^{-1} times the characteristic function of $[-\delta, 0]$. Then

$$o(1) = \int_x^{x+\delta} \delta^{-1}\phi(y)\,dy \leq \delta^{-1}\int_x^{x+\delta} (\phi(x) + 2\delta)\,dy = \phi(x) + 2\delta ,$$

so that $\underline{\lim}\ \phi(x) \geq 0$. If we take the characteristic function of $[0, \delta]$ for δH we get the opposite inequality, and it is readily seen that the appropriate tauberian condition can be formulated

$$(3) \qquad \lim_{h\downarrow 0}\ \overline{\lim_{x\to\infty}}\ \sup_{x\leq y\leq x+h} (\phi(y) - \phi(x)) = 0.$$

That (1) and (3) imply $\phi(x) = o(1)$, $x\to\infty$, is usually called Wiener-Pitt's theorem. We have just proved that Wiener's tauberian theorem on page 1 implies Hardy-Littlewood's theorem. If we apply our transformations to the remainder theorem considered on page 2, we see that that theorem can be rephrased in the following way:

(4) $\qquad K * \phi(x) = O(\exp(-\delta x)), \quad x\to\infty,$

<u>and</u>

(5) $\qquad \phi(y) - \phi(x) \leq O(x^{-1}), \quad x \leq y \leq x + x^{-1}, \quad x\to\infty,$

<u>imply</u>

$$\phi(x) = O(1/x)$$

(condition (5) is the appropriate weaker form of (1.1.4) but requires more than (3)). Formula (4) follows since

$$(1-x)^{\delta} = (1-e^{-t})^{\delta} = O(t^{\delta}) = O(e^{-\delta u})$$

if $x = \exp(-t)$, $t = \exp(-u)$.

In order to see that the conclusion should be $\phi(x) = O(1/x)$ we observe that $\sum_{n\leq x} a_n = \alpha(x) = \phi(\log x)$.

The theorem is not true without further conditions on K, one natural condition being that <u>there is a function</u> g <u>holomorphic in a strip around the real axis and fulfilling</u>

(6) $\qquad g(t)\hat{K}(t) = 1$ <u>for real</u> t,

<u>and</u>

(7) $\qquad |g(t)|^{-1} \leq A\exp(a|t|)$ <u>for</u> t <u>in the strip</u>.

(A more general formulation is given in theorem 2.1.)

This general tauberian remainder theorem was deduced in Ganelius 1958 from the result for the Laplace transform. In 1962 I gave a direct proof using Fourier transforms, which is the method recommended by Beurling, who in 1938 stated the first general tauberian remainder theorem. Beurling considered another class of kernels for which the bound in (7) is not exponential but polynomial. Beurling's work has been continued by Lyttkens in 1954, 1956 and in some unpublished manuscripts. In our next chapters we shall consider this type of remainder problems.

Before giving the proof of Wiener's theorem which will be the model for my proof of the remainder theorems, I recall some equivalent formulations of the theorem (cf. Pollard 1953).

WIENER'S APPROXIMATION THEOREM:

The linear subspace spanned by the translates $\{K(.-\lambda)\}_{\lambda\in R}$ of any $K \in L(-\infty,\infty)$ is dense in $L(-\infty,\infty)$ if and only if $\hat{K}(t) \neq 0$ for all real t.

The 'only if' part of the theorem is trivial, since every function in the closure must have its Fourier transform vanishing at a zero of K. The conclusion of the 'if' part of the theorem says that if $H \in L$, then to every $\varepsilon > 0$ there are sets $\{a_n\}$ and $\{\lambda_m\}$ so that

$$||H - \sum_1^n a_m K(.-\lambda_m)|| = \int |H(x) - \sum_1^n a_m K(x-\lambda_m)|dx < \varepsilon .$$

Hence

$$|H * \phi(x)| \leq \varepsilon ||\phi||_\infty + \sum_1^n |a_m| \, |K*\phi(x-\lambda_m)|,$$

and $\lim_{x\to\infty} K*\phi(x) = 0$ evidently implies that $\lim_{x\to\infty} H*\phi(x) = 0$. Thus the general tauberian theorem follows from this approximation theorem.

We next show that the general tauberian theorem implies the following result. (Carleman 1944, p.115, cf. also Beurling 1945).

If $K \in L(-\infty,\infty)$ and $\hat{K}(t) \neq 0$ for all real t, then $K*\phi = 0$ implies $\phi = 0$.

Assume that $\phi_0 \neq 0$ is a solution of $K*\phi_0(x) = 0$. We may without loss of generality assume that ϕ is uniformly continuous. (Let J be continuous with compact support and look at $J*\phi$. Cf. Beurling 1945). According to Wiener-Pitt's theorem $\lim_{|x|\to\infty} \phi_0(x) = 0$.

We can achieve $\phi_0(0) = 1$ by multiplication by a suitably number after a translation. Determine a convex sequence $(\lambda_n)_1^\infty$ such that $|\phi_0(x)| < 4^{-n}$ for $|x| > \lambda_n - \lambda_{n-1}$ and form $\phi(x) = \sum_1^\infty \phi_0(x-\lambda_n)$. It is easy to see that the series converges and that the boundedness of ϕ follows from that of ϕ_0. But evidently $\overline{\lim}_{n\to\infty} |\phi(\lambda_n)| > 0$ and $K*\phi = 0$ (since it holds for every partial sum). Thus we get a contradiction to Wiener's general theorem. Hence $\phi_0 = 0$.

That the theorem just derived from Wiener's general theorem implies the approximation theorem follows by Hahn-Banach's theorem, the functionals on L being given by the L^{∞}-functions. Assume that the closure of subspaces spanned by the translates is not the whole of L. Then, according to a corollary to Hahn-Banach's theorem (cf. e.g. Rudin 5.19) there is a $\phi_0 \in L^{\infty}$ such that $\phi_0 \neq 0$ but $\int K(\lambda - x)\phi_0(x)dx = 0$ for all λ. Hence all the three theorems are equivalent.

It is known (cf. Ganelius 1969) that general remainder theorems have associated approximation theorems giving more precise information than Wiener's approximation theorem, but of course only for restricted classes of kernels.

Let us now turn to the proof of Wiener's theorem. Two functions in L are given, H and K. The fundamental idea is to define two functions U and V in L such that

$$H * \phi = U * \phi + V * (K * \phi) \tag{8}$$

If this construction is possible with an U of arbitrarity small norm, Wiener's theorem is proved, since it follows that

$$\overline{\lim_{x \to \infty}} |H * \phi(x)| \leq \|\phi\|_{\infty} \|U\| + \overline{\lim_{x \to \infty}} |V * (K * \phi)(x)|$$

and the last term on the right is zero by Lebesgue's convergence theorem.

For the construction of U and V we need two lemmas, Wiener's lemma on page 2 and the following.

<u>**LEMMA:**</u>

<u>To every</u> $F \in L(-\infty, \infty)$ <u>and</u> $\varepsilon > 0$, <u>there is an</u> $F_* \in L(-\infty,\infty)$ <u>with</u> $\|F - F_*\| < \varepsilon$, <u>and compact support of</u> $\hat{F}_*$.

To get a proof of this lemma we just write down an F_* with the property mentioned.

Let h be a positive function with $\int h(x)dx = 1$ and with a Fourier transform with support in $[-1,1]$. We can for instance choose $h(x) = (\pi x)^{-2}\sin^2 \pi x$.

We define F_ω by

$$F_\omega(x) = \int F(x-y)h(\omega y)\,\omega\, dy,$$

and observe that $\hat{F}_\omega(t) = \hat{h}(\frac{t}{\omega})\hat{F}(t)$. Hence the support is in $[-\omega, \omega]$. Now

$$\begin{aligned} \| F-F_\omega \| &= \int \left| F(x) - \int F(x-y)\,\omega h(\omega y)dy \right| dx \\ &= \int \left| \int (F(x) - F(x-y))\,\omega h(\omega y)dy \right| dx \qquad (9) \\ &\leq \int h(y) \int |F(x) - F(x-y\omega^{-1})|\,dx\,dy. \end{aligned}$$

Now it is a well-known result that $F \in L$ implies that

$$\lim_{h \to 0} \int |F(x) - F(x-h)|\,dx = 0$$

and since $\int |F(x) - F(x-y\omega^{-1})|\,dx \leq 2\|F\|$, Lebesgue's convergence theorem applied to the last line in (9) proves that

$$\lim_{\omega \to \infty} \| F-F_\omega \| = 0.$$

Hence we can always find an ω_1 so that

$$\| F-F_{\omega_1} \| < \varepsilon \quad \text{and take} \quad F_* = F_{\omega_1}.$$

Returning to (8) we choose $U = H-H_*$, where $\hat{H}_*$ has compact support and $\| H-H_* \| < \varepsilon$. If we can prove that there is a $V \in L$ with $K*V = H_\omega$, then it holds that

$$H = U + V * K,$$

and we are done, since the proof following formula (8) works.

Now according to Wiener's lemma $(\hat{K})^{-1}$ is locally a Fourier transform. Let us cover the compact support of $\hat{H}_*$ with open sets $\{O_\tau\}$ in which $(\hat{K})^{-1}$ coincides with the Fourier transform of k_τ. By the compactness it is enough to take a finite number of sets $\{O_\tau\}_{\tau\in\Lambda}$. Let $1 = \sum h_\tau$ be a partition of unity, subordinate to $\{O_\tau\}$ and with twice differentiable h_τ. Then

(10) $$\hat{K}(t)^{-1} = \sum_{\tau\in\Lambda} h_\tau(t)\, k_\tau(t) \quad \text{for} \quad t \in \bigcup O_\tau$$

and in particular for all t in the support of $\hat{H}_*$. Now the sum on the right in (10) is evidently in the set $\hat{L}$ of Fourier transforms of L-functions (since multiplication of transforms corresponds to convolution of L-functions). Thus $\hat{H}_*(\hat{K}(t))^{-1}$ coincides on supp $\hat{H}_*$ with the Fourier transform of a function $V \in L$.

Thus everything is brought back to Wiener's lemma, and we conclude this section with a proof, which in fact yields the slighty stronger result (due to Lévy 1934) that if G is holomorphic at $\hat{K}(\tau)$ then $G(\hat{K})$ coincides with the

Fourier transform of an integrable function in a neighborhood of τ .

Let us put $\hat{K}(\tau) = z_0$. Consider the Taylor series of G around z_0

$$G(z) = \sum_0^{\infty} c_k (z-z_0)^k$$

and let the radius of convergence be ρ . Since $\hat{K}$ is continuous, there is an open interval I around τ such that $|\hat{K}(t) - z_0| < \rho$ for $t \in I$.

Let $h \in L$ and suppose that $\operatorname{supp} \hat{h}(t) \subset [-1,1]$ and that $\hat{h}(t) = 1$ for all t in a neighborhood of 0. Put $\phi(t) = \hat{h}(\varepsilon^{-1}(t-\tau))$, and consider the formula

$$G(\hat{K}(t)) = \sum_1^{\infty} c_k(\hat{K}(t)\phi(t) - z_0\phi(t))^k + c_0\phi(t), \tag{11}$$

which according to our constructions holds in a neighborhood of τ depending on ε .

Every term on the right belongs to $\hat{L}$, since

$$(\hat{K}(t)\phi(t) - z_0\phi(t))^k = (\underbrace{u * u * \ldots * u}_{k \text{ times}})^\wedge$$

where

$$u(x) = \int K(y)e^{2\pi i\tau(x-y)} \varepsilon h(\varepsilon(x-y))dy - e^{2\pi i\tau x} \varepsilon h(\varepsilon x)\int e^{-2\pi i y\tau} K(y)dy. \tag{12}$$

If we can choose ε so that $\|u\| < \rho$, then the completeness of L will guarantee that the series on the right of (11) is in $\hat{L}$. So it only remains to show that $\lim_{\varepsilon\to 0} \|u\| = 0$.

Formula (12) may be rewritten

$$u(x) = \int K(y)e^{2\pi i\tau(x-y)}(h(\varepsilon(x-y)) - h(\varepsilon x))\varepsilon\, dy,$$

and hence

$$\|u\| \le \int |K(y)| \int |h(\varepsilon(x-y)) - h(\varepsilon x)|\varepsilon\, dx\, dy =$$

$$= \int |K(y)| \int |h(s-\varepsilon y) - h(s)| ds\, dy.$$

We now have exactly the same expression as in the last member of (9), and by the same arguments it follows that $\|u\| \to 0$ as $\varepsilon \to 0$.

Taking a sufficienly small ε , we get $\|u\| < \rho/2$, and the integrable functions $s_n(x) = c_0 e^{2\pi i\tau x}\varepsilon h(\varepsilon x) + \sum_1^n c_k u^{*k}$, which have the partial sums of the series on the right of (11) as Fourier transforms, form a Cauchy sequence, since

$$\|s_n - s_m\| \le \sum_{m+1}^{n} |c_k|(\rho/2)^k.$$

The limit-function s fulfills

$$\hat{s}(t) = G(\hat{K}(t))$$

in a neighborhood of τ .

The proof of Wiener's theorems we have given is a classical hard analysis proof. I have not at all pursued the line I mentioned in the introduction, i.e. the approach given by Gelfand. The main reason is that I don't know of any interesting remainder result on any more complicated group than R^d. Even in cases like the integers Z and the circle T it might be a matter of discussion, if the results qualify as tauberian remainder theorems (for T see Theorem 2 in Ganelius 1957a and for some comments on Z see section 8.3).

(All the proofs given here can immediately be applied in R^d, if $x \to \infty$ is interpreted as $\min_{1 \leq m \leq d} |x_m| \to \infty$. Some of the problems in d dimensions are discussed by Frennemo 1966, 1967).

1.4 The theory of distributions

There are some rather simple proofs of Wiener's theorem which rely on the fundamental ideas of the theory of distributions (e.g. Korevaar 1965). However, Wiener's theorem does not seem to be the natural theorem, if we work within that theory. All the classical tauberian theorems also follow from the following theorem, which I have not seen stated but which is implicit in standard treatments of Fourier transforms of distributions (Schwartz 1949, 1960; Friedman 1959). The main reason for including this theorem here is that the method will be applied in a later chapter.

We use the customary notations. $\mathcal{S}$ stands for the rapidly decreasing C^∞-functions, $\mathcal{S}'$ for the tempered distributions, O_M for the slowly increasing C^∞-functions and O_C' for the rapidly decreasing distributions.

Every function which does not increase more rapidly than a polynomial belongs to $\mathcal{S}'$ and every function decreasing to zero more rapidly than any polynomial belongs to O_C' . Linear combinations of translates and derivatives of δ-distributions belong to O_C' .

For tempered distributions we write $T = o(1)$, if for every $\varphi \in \mathcal{S}$, it holds that $T * \varphi(x) = o(1)$, $x \to \infty$. As usual $\mathcal{D}$ denotes the set of test functions, i.e. C^∞-functions with compact support. The set of distribution is called $\mathcal{D}'$. The Fourier transforms of distributions in O_C' are functions in O_M. It might be pointed out that if T is a function f , then $T = o(1)$ is implied by $f(x) = o(1)$, $x \to \infty$.

A TAUBERIAN THEOREM FOR DISTRIBUTIONS:

Let $K \in O_C'$, $\phi \in \mathcal{S}'$ *and suppose that* $\hat{K}(t) \neq 0$ *for all real* t. *Then*

$K * \phi = o(1)$, <u>implies that</u> $H * \phi = o(1)$ <u>for every</u> $H \in O_C'$. <u>In particular</u> $\phi = o(1)$.

To prove this theorem we take a $\theta \in \mathcal{D}$, and consider

$$(H*\phi)*\hat{\theta}(x) = \hat{H}\hat{\phi}(\theta \exp(2\pi ix.)) = \hat{\phi}(\hat{H}\theta \exp(2\pi ix.)).$$

The first equality holds, since under the conditions given $(H*\phi)^\wedge = \hat{H}\hat{\phi}$ and the second is the rule for multiplication of a distribution by a C^∞-function. If $\psi = K*\phi$, then $\psi \in \mathcal{S}'$ and

$$\hat{\psi} = \hat{K}\hat{\phi}.$$

Hence

$$(H*\phi)*\hat{\theta}(x) = \hat{K}^{-1}\hat{\psi}(\hat{H}\theta \exp(2\pi ix.)) = \hat{\psi}((\hat{H}/\hat{K})\theta \exp(2\pi ix.)) = \psi * \chi(x)$$

Where $\hat{\chi} = (\hat{H}/\hat{K})\ \theta \in \mathcal{D}$ and thus $\chi \in \mathcal{S}$.

It follows that $\psi * \chi(x) = o(1)$ and hence

$$(H*\phi)*\chi(x) = o(1)$$

for all $\chi \in \hat{\mathcal{D}}$. But $\hat{\mathcal{D}}$ is dense in $\mathcal{S}$, and hence

$$(H*\phi)*\chi(x) = o(1) \quad \text{for all } \chi \in \mathcal{S},$$

and it follows that

$$H*\phi = o(1).$$

In particular we can take $H = \delta$, and get $\phi = o(1)$.

This theorem requires more of K and less of ϕ than Wiener's theorem. The point I want to make is that all the classical kernels are in O_C' and hence for each special kernel we have a more general result, since we need not assume that ϕ is bounded, only tempered.

As an example we observe the following proof of Littlewood's theorem that $\sum_0^\infty a_n x^n \to 0$, $x \uparrow 1$ and $na_n = O(1)$ imply $\sum_0^\infty a_n = 0$. The transformations we made on page 7 give us $K * \phi(x) = o(1)$, where $K(x) = \exp(-x - \exp(-x))$, which evidently belongs to O_C' and $\phi(x) = \sum_{n<\exp x} a_n = O(1) \sum_{n<\exp x} n^{-1} = O(x)$ so that $\phi \in \mathcal{S}'$. Our distribution tauberian theorem gives $\int \phi(x-y)\ \theta(y)dy \to 0$ for every $\theta \in \mathcal{S}$ and hence by the tauberian condition it follows as above that $\phi(x) = o(1)$, $x \to \infty$.

The methods we have introduced in this section will be applied in the proof of some uniqueness theorems in Chapter 6. (For other similar variants of

Wiener's theorem, cf.Stenström 1961).

1.5 Examples of kernels of specific interest

In order to justify my claim that almost all kernels in the classical tauberian theorems are rapidly decreasing, I give a collection of different examples (cf. also Hardy 1949, p.297-8, Hirschman-Widder 1955, p.78-79).

1. The Laplace transform

$$K_1(x) = \exp(-px - \exp(-x)), \quad \hat{K}_1(t) = \Gamma(p + 2\pi it), \quad \text{where} \quad p > 0.$$

In our treatment $p = 1$, but it is often practical to use different p:s depending on the tauberian condition, so that all such can be reduced to the form $|\phi'(x)| \leq 1$, or more generally to $\phi(y) - \phi(x) \leq c\, s(x)$ for all $x_0 \leq x \leq y < x + s(x)$ and a certain $s(x)$ (cf. Frennemo 1965).

2. The Stieltjes transform

$$K_2(x) = (1 + \exp(-x))^{-1-\rho}\exp(-(\nu + 1)x), \quad \rho > \nu \geq 0,$$

$$\hat{K}_2(t) = B(\nu + 1 + 2\pi it, \rho - \nu - 2\pi it) =$$

$$= \Gamma(\nu + 1 + 2\pi it)\Gamma(\rho - \nu - 2\pi it)/\Gamma(\rho)$$

(cf. Ganelius 1964).

3. The kernel of Cesàro and Riesz summation

$$K_3(x) = \begin{cases} \Gamma(p)^{-1}(1 - \exp(-x))^{p-1}\exp(-x), & \text{if} \quad x > 0 \\ 0 \quad \text{if } x \leq 0 \end{cases}$$

$$\hat{K}_3(t) = \Gamma(1 + 2\pi it)/\Gamma(1 + p + 2\pi it), \quad \operatorname{Re} p > 0$$

(cf. Ganelius 1962).

4. A Bessel kernel

$$K_4(x) = J_r(\exp(-x)) \exp(-qx), \quad q + r > 0, \quad q < \tfrac{1}{2},$$

$$\hat{K}_4(t) = 2^{q-1+2\pi it}\Gamma(\tfrac{1}{2}(q + r + 2\pi it))/\Gamma(1 + \tfrac{1}{2}(r - q - 2\pi it)).$$

(cf. Bochner-Chandrasekharan 1949, p.67, Bureau 1960, 1962, Ganelius 1962).

5. Weierstrass's kernel

$$K_5(x) = \exp(-\pi x^2),$$

$$\hat{K}_5(t) = \exp(-\pi t^2).$$

6. $K_6(x) = \exp(-c|x|), \quad c > 0,$

$$\hat{K}_6(t) = 2c(c^2 + 4\pi^2 t^2)^{-1}$$

(for the corresponding kernel in d dimensions, see Bochner-Chandrasekharan p.69).

7. The Meijer transform

$$K_7(x) = e^x K_\nu(e^x), \quad -1 < \nu < 1. \qquad (K_\nu \text{ modified Bessel function}).$$

$$\hat{K}_7(t) = 2^{-1-2\pi i t}\Gamma(\tfrac{1}{2}(1 + \nu - 2\pi i t)\Gamma(\tfrac{1}{2}(1 - \nu - 2\pi i t)))$$

(cf. Hirschman-Widder 1955, p.78).

8. A kernel related to the prime number theorem

$$K_8(x) = (2[\exp(-x)] - a[a^{-1}\exp(-x)] - b[b^{-1}\exp(-x)])e^x,$$

(a and b positive, $\log a/\log b$ irrational)

$$\hat{K}_8(t) = (1 - 2\pi i t)^{-1}\zeta(1 - 2\pi i t)(2 - \exp(2\pi i t \log a) - \exp(2\pi i t \log b))$$

(cf. Hardy 1949, p.298 ff., Ingham 1945).

9. A kernel of interest in remainder theory

$$K_9(x) = \exp(-\beta x)\sin(\exp x), \quad 0 < \beta < 1.$$

$$\hat{K}_9(t) = -\Gamma(-\beta - 2\pi i t)\sin\frac{\pi}{2}(\beta + 2\pi i t)$$

(cf. 4.3.Ex.3 and Titchmarsh 1948, p.204).

10. A kernel connecting minimum modulus and zerodistribution for functions of order less than one

$$K_{10}(x) = K_{11}(x) - \int_x^\infty K_{11}(y)\,dy$$

$$\hat{K}_{10}(t) = (2it)^{-1}(\cos\pi\mu - \cos\pi\lambda)/\sin\pi\mu, \quad \mu = \lambda + 2\pi i t, \quad 0 < \lambda < 1,$$

(cf. Essén-Ganelius 1967, and for $\lambda = \frac{1}{2}$ Beurling 1938, p.22), where

$$K_{11}(x) = \exp(-\lambda x)(\log|1 - \exp x| - \cos \pi\lambda \quad \log|1 + \exp x|)$$

$$\hat{K}_{11}(t) = \frac{\pi}{\mu}(\cos \pi\mu - \cos \pi\lambda)/\sin \pi\mu , \qquad \mu = \lambda + it, \quad 0 < \lambda < 1.$$

11. The kernel in Lambert summation

$$K_{12}(x) = p(\exp(-x)) \exp(-x)$$

with

$$p(u) = - \frac{d}{du}\left(\frac{u}{\exp u - 1}\right),$$

$$\hat{K}_{12}(t) = 2\pi it \, \zeta(1 + 2\pi it)\Gamma(1 + 2\pi it)$$

(cf. Wiener 1933, p. 112 ff., Pitt 1958, p. 80 ff.).

CHAPTER 2

A TYPICAL GENERAL REMAINDER THEOREM

2.1 A general remainder theorem generalizing Hardy-Littlewood's theorem

In this chapter we shall give a concise proof of a remainder theorem which is a generalization of the main results for Laplace transforms obtained by Freud and Korevaar. The method was first applied in Ganelius 1962. Sub - additivity of a function $f: R \to R_+$ means that $f(x+y) \leq f(x) + f(y)$.

THEOREM 2.1 :

Let $K \in L(R)$ *and assume that there is a function* g *holomorphic in a strip around the real axis and such that*

$$\hat{K}(t) g(t) = 1 \quad \text{for all real } t,$$

(1) $$|g(t)| \leq A \exp(a|t|) \quad \text{for } t \text{ in the strip.}$$

Let $\phi \in L^{\infty}(R)$ *and assume that*

$$\psi(x) = K * \phi(x) = O(\exp(-W(x))),$$

with a positive increasing sub-additive W *such that*

(2) $$\eta = -(2\pi i)^{-1} \lim_{x \to \infty} x^{-1} W(x)$$

belongs to the strip mentioned above. If ϕ *satisfies the tauberian conditions*

(3) $$\sup_{x \leq y \leq x + W(x)^{-1}} (\phi(y) - \phi(x)) \leq O(1/W(x)), \quad x \to \infty$$

then

(4) $$\phi(x) = O(1/W(x)), \quad x \to \infty .$$

It will follow from the theorems of next chapter, that this theorem is best possible in the sense, that no function tending to zero can be substituted for 1 on the right in (4) if $\hat{K}(t) = O(\exp(-b|t|))$ in the strip. In that chapter we also show how to obtain estimates for means of ϕ.

2.2 A basic formula.

Let K and ϕ fulfill the conditions of theorem 2.1 and put $\psi = K * \phi$. We shall first prove that under weak conditions on the functions $E: R \to R_+$ and $\chi: R \to R_+$ it holds that

(1) $$\int \phi(x-y)E(y)\chi(y-\eta)\, dy = Q*\psi(x) \quad \text{for all} \quad \eta \in R,$$

where

(2) $$\hat{Q}(t) = \hat{K}(t)^{-1} \int \hat{E}(t-u)\hat{\chi}(u)e^{-2\pi i u\eta}\, du.$$

We shall apply similar formulas in several of the following chapters. Let us for the moment prove the statement in the case that

$$E(y) = \exp(-\pi^2\omega^{-1}y^2), \qquad \chi(y) = \Omega(\pi y\Omega)^{-2}(\sin \pi y\Omega)^2$$

with positive ω and Ω.

Then
$$\hat{E}(u) = \exp(-\omega u^2)\left(\frac{\omega}{\pi}\right)^{1/2}$$
$$\hat{\chi}(u) = 1 - \Omega^{-1}|u|$$

for $|u| < \Omega$ and $= 0$ elsewhere.

Since $\hat{\chi}$ has compact support, the integral in (2) gives an entire function in t and that function decreases like $\exp(-\omega t^2)$ if $|\operatorname{Re} t| \to \infty$. Hence $\hat{Q}$ is holomorphic in the same strip as $g = \hat{K}^{-1}$ and it belongs to L on every horizontal line in that strip. It follows that $Q \in L$ and an application of Fubini's theorem yields

$$Q * \psi = (Q * K) * \phi = H * \phi$$

with $H \in L$ and

$$\hat{H}(t) = \int \hat{E}(t-u)\hat{\chi}(u)e^{-2\pi i u\eta}\, du = \int e^{-2\pi i t y}E(y)\chi(y-\eta)\, dy$$

where the last expression follows by Parseval's formula. Now formula (1) is an immediate consequence.

2.3 Proof of theorem 2.1

To get an estimate for $|\phi(x)|$ we next introduce

$$M_x = \sup_y |\phi(x-y)E(y)| \geq |\phi(x)E(0)| = |\phi(x)|$$

and chose η so that

$$\phi(x-\eta-\Omega^{-1})E(\eta+\Omega^{-1}) > \tfrac{3}{4}M_x$$

(We may without loss of generality assume that the extreme value of $\phi(x-y)E(y)$ is positive.)

If we check that

$$\int_{\Omega|w|>1} \chi(w)\, dw < \frac{1}{3}, \qquad \int_{\Omega|w|<1} \chi(w)\, dw > \frac{2}{3},$$

it follows from (2.2.1) that

$$\frac{2}{3}\left\{\frac{3}{4}M_x - \sup_{\eta-\Omega^{-1} \leq y \leq \eta+\Omega^{-1}} (\phi(x-y)E(y) - \phi(x-\eta-\Omega^{-1})E(\eta+\Omega^{-1}))\right\} - \frac{1}{3}M_x \leq$$

$\leq |Q*\psi(x)|$.

Hence

(1) $$|\phi(x)| \leq M_x \leq 4 \sup_{0 \leq v-y \leq 2\Omega^{-1}} (\phi(x-y)E(y) - \phi(x-v)E(v)) + 6\,|Q*\psi(x)|.$$

The first term on the right will be estimated by aid of the tauberian condition. Let us start with the second and observe that according to (2.2.2) we have

$$|Q(x)| = \left| \int e^{2\pi itx} g(t) \int \hat{E}(t-u)\, \hat{\chi}(u) e^{-2\pi iu\eta}\, du\, dt \right|.$$

We take $\omega = \pi$ so that $\hat{E}(u) = \exp(-\pi u^2)$. If $i\delta$ belongs to the strip of holomorphy of g, then by the transformation $t \to t+i\delta$ we get

$$|Q(x)| \leq A \int \exp(-2\pi\delta x + a|t|) \int_{|u|\leq\Omega} \exp(\pi\delta^2 - \pi(t-u)^2)\, du\, dt.$$

Inverting the order of integration and substituting $t=u+v$ we find

$$Q(x) = O(1)\exp(-2\pi\delta x) \int_{|u|\leq\Omega} du \int \exp(a|u| + a|v| - \pi v^2)\, dv =$$

$$= O(1)\exp(a\Omega - 2\pi\delta x).$$

With $i\delta$ replaced by $-i\varepsilon$ we obtain $Q(x) = O(1)\exp(a\Omega + 2\pi\varepsilon x)$.
Hence

$$Q*\psi(x) = O(1)\exp(a\Omega)\left\{ \int_{-\infty}^{0} \exp(2\pi\varepsilon y - W(x-y))\,dy + \int_{0}^{\infty} \exp(-2\pi\delta y - W(x-y))\,dy \right\}$$

and the sub-additivity of W gives $-W(x-y) \leq W(y) - W(x)$, so that

(2) $$Q*\psi(x) = O(1)\exp(a\Omega - W(x))\left\{ \int_{-\infty}^{0} \exp(2\pi\varepsilon y + W(y))\,dy + \int_{0}^{\infty} \exp(-2\pi\delta y + W(y))\,dy \right\}.$$

Both integrals are bounded, the first since W is bounded for $y<0$, and the second on account of (2.1.2) if δ is properly chosen.

We are going to take $\Omega = (2a)^{-1}W(x)$ and if we can prove that the first term on the right in (1) is $O(\Omega^{-1})$ our theorem follows, since by (2) we see that $Q*\psi(x) = O(1)\exp(-\frac{1}{2}W(x)) = O(1/W(x))$. Now

(3) $$\phi(x-y)E(y) - \phi(x-v)E(v) = (\phi(x-y)-\phi(x-v))E(y) + (E(y)-E(v))\phi(x-v).$$

If $|y| > \frac{1}{2}x$, then all the terms on the right are $O(1)\exp(-\pi x^2/4) = O(1/W(x))$, since every sub-additive function is bounded by a linear function.

Let us assume that $|y| \leq \frac{1}{2} x$, that $2a\Omega = W(x)$ and that x is sufficiently large. Since E' and Φ are bounded the second term on the right in (2.3.3) is $O(v-y) = O(\Omega^{-1}) = O(1/W(x))$. The same estimate holds for the first by (2.1.3) and the observation that $W(x) \leq 2W(y) \leq 2W(x)$, if $\frac{x}{2} \leq y \leq x$.

Hence we have obtained estimates of the proper order $O(1/W(x))$ for all terms in (1), and theorem 2.1 is proved.

CHAPTER 3

REMAINDER THEOREMS IN WIENER'S FORM

3.1 Introduction. Notation and the fundamental formula

In the previous chapter we have given a simple and straightforward proof of a typical general remainder theorem. It is probably not easy to see the close connection between that proof and the proof of Wiener's theorem on p.10. There is however such a connection.

In this chapter we are going to modify the method applied on p.10 in such a way that it gives interesting results for the general tauberian remainder problem stated in the introduction (p.3). Some of the formulas we use are exactly the same as were considered in the preceding chapter. The generality of the tauberian condition in theorem 2.1 makes the res emblance less clear. However, if we strengthen that condition to $|\phi'| \leq 1$, then the proof of the theorem is an easy application of the methods of this chapter as will be shown in section 3.3. We also prove that theorem 2.1 is essentially best possible.

The main purpose of this chapter is to show that very small modifications of the natural method for proving Wiener's theorem give interesting remainder results. Some unnecessary restrictions are introduced in theorems 3.2.1 and 3.2.2 to avoid complications in the proof.

Theorems 3.2.1 and 3.2.2 and methods to prove them have not been published before. Except the connections of these theorems with theorem 2.1 and hence with previous remainder theorems for Laplace and Stieltjes transforms, I want to mention the close relation of corollary 3.2.3 to some convexity results of Chandrasekharan-Minakshisundaram. We are in fact dealing with a kind of interpolation problem, and the methods are related to those applied in the theory of interpolation spaces (see e.g. Peetre 1964).

We are going to modify the formula

(1) $$H * \phi = U * \phi + V * \psi, \qquad \psi = K * \phi$$

which we discussed on p.10. The following notation will be used in our treatment of the case of $d \geq 1$ dimensions. The measurable essentially bounded functions, the measurable functions with integrable $|f|^p$ and the k times continuously differentiable functions are as usual denoted by $L^\infty(R^d)$, $L^p(R^d)$ and $C^k(R^d)$. If suitable we omit R^d and if integral signs without limit occur they refer to R^d. The L^p norm of $f \in L^p(R^d)$ is denoted $\|f\|_p$.

If $(x_1, x_2, \ldots, x_d) = x \in R^d$ and $(y_1, y_2, \ldots, y_d) = y \in R^d$ then $x+y = (x_1+y_1, x_2+y_2, \ldots, x_d+y_d)$ and $x<y$ means $x_k<y_k$ for $1 \leq k \leq d$.

For the norm we write

$$||x|| = (\sum_1^d |x_k|^2)^{1/2}$$

whereas

$$|x| = (|x_1|, |x_2|, \ldots, |x_d|)$$

We shall use $x \to \infty$ for $x_k \to \infty$ for all k so that $|x| \to \infty$ means $|x_k| \to \infty$ for all k. Sometimes we shall consider $\mathbb{C}^d$, and if

$$z = (z_1, z_2, \ldots, z_d) \in \mathbb{C}^d,$$

we write as above

$$|z| = (|z_1|, |z_2|, \ldots, |z_d|)$$

but also

$$\text{Re } z = (\text{Re } z_1, \text{Re } z_2, \ldots, \text{Re } z_d),$$

and correspodingly

$$\text{Im } z = (\text{Im } z_1, \text{Im } z_2, \ldots, \text{Im } z_d).$$

The scalar product of $x, y \in R^d$ is written

$$xy = \sum_1^d x_k y_k$$

and hence

$$x^2 = \sum_1^d |x_k|^2 = ||x||^2 .$$

The Fourier transform $\hat{f}$ of $f \in L(R^d)$ is defined by

$$\hat{f}(t) = \int \exp(-2\pi i t x) f(x)\, dx =$$
$$= \int \exp(-2\pi i \sum t_k x_k) f(x_1, x_2, \ldots, x_d)\, dx_1 dx_2 \ldots dx_d$$

A function $f: R^d \to R_+$ is called sub-additive if for all $x, y \in R^d$, it holds that

$$f(x+y) \leq f(x) + f(y) .$$

Examples of sub-additive functions are

$$f_1(x) = ||x||$$

and

$$f_2(x) = m \log(1+||x||), \quad m > 0.$$

It is evident that every sub-additive function is dominated by some linear expression

$$a + b|x| = a + \sum_1^d b_k|x_k|$$

where $a \in R$ and $b \in R_+^d$.

In order to formulate the modification of (1) we next introduce a function $\gamma \in L^\infty(R^d)$ with

$$\gamma(x) = \begin{cases} 0 & \text{for } ||x|| \geq 2 \\ 1 & \text{for } ||x|| \leq 1 . \end{cases}$$

If $H \in L(R^d)$ and $K \in L(R^d)$ we define U and T by

(2) $$\hat{U}(t) = \hat{H}(t) \int (1-\gamma(u/\Omega)) \exp(-\omega(u-t)^2)(\omega/\pi)^{d/2}\, du$$

and

(3) $$\hat{T}(t) = (\hat{H}(t)/\hat{K}(t)) \int \gamma(u/\Omega) \exp(-\omega(u-t)^2)(\omega/\pi)^{d/2}\, du ,$$

where ω and Ω are positive numbers. If we impose a suitable restriction on the growth of $\hat{K}$ at infinity we see that U and T belong to $L(R^d)$, so that for every pair $\phi, \psi \in L^\infty(R^d)$ with $\psi = K * \phi$ it holds that

(4) $$H * \phi = U * \phi + T * \psi .$$

This formula is a consequence of

$$\hat{H} = \hat{U} + \hat{K}\hat{T} ,$$

which evidently follows from (2) and (3), if U and $T \in L(R^d)$, since

$$\int \exp(-\omega(u-t)^2)\, du = (\pi/\omega)^{d/2} .$$

For certain purposes it might be suitable to consider the more general formula with

$$\sum_1^d \omega_k(u_k-t_k)^2 \quad \text{instead of} \quad \omega(u-t)^2 = \sum_1^d \omega(u_k-t_k)^2 ,$$

but we are not now going to pursue the matter in that direction (cf. Chapter 5).

We are now ready to formulate a general remainder theorem and also a kind of converse theorem showing the precision of the first mentioned theorem.

3.2 A general remainder theorem and its converse

If $\sigma \in R_+^d$ we define the subset D_σ of $\mathbb{C}^d$ by

$$D_\sigma = \{t \in \mathbb{C}^d : |\mathrm{Im}\ t| \le \sigma\}.$$

THEOREM 3.2.1 :

Let $H \in L(R^d)$. Suppose that there is an η, holomorphic in D, and a sub-additive h such that

$$\eta(t) = \hat{H}(t) \quad \text{for} \quad t \in R^d,$$

and

(1) $$|\eta(t)| \le \exp(-h(\mathrm{Re}\ t)) \quad \text{for} \quad t \in D_\sigma .$$

Let $K \in L(R^d)$ and suppose that there is a $\varkappa$, holomorphic in D_ε, $\delta > \varepsilon$ and a sub-additve k such that

$$\hat{K}(t)\varkappa(t) = 1 \quad \text{for} \quad t \in R^d,$$

and

(2) $$|\varkappa(t)| \le \exp(k(\mathrm{Re}\ t)) \quad \text{for} \quad t \in D_\varepsilon .$$

Let $\phi \in L^\infty(R^d)$, put $\psi = K * \phi$ and assume that

(3) $$\phi(x) = O(\exp(-V(x))), \quad x \to \infty,$$

(4) $$\psi(x) = O(\exp(-W(x))), \quad x \to \infty,$$

with positive sub-additive V and W satisfying

(5) $$\int \exp(V(x) - 2\pi\delta\,|x|)\,dx < \infty, \quad \int \exp(W(x) - 2\pi\varepsilon\,|x|)\,dx < \infty .$$

Then it holds that

(6) $$H * \phi(x) = O(1) \inf_{\Omega > 0} \Big\{ \exp(-V(x)) \int (1 - \gamma(t/\Omega)) \exp(-h(t))\,dt + \exp(-W(x)) \int \gamma(t/\Omega) \exp(k(t) - h(t))\,dt \Big\}.$$

That the estimate given in the above theorem is precise in many cases is shown by the following result.

THEOREM 3.2.2 :

Let $H, K \in L(R^d)$ *be such that there are functions* η *and* $\varkappa$, *holomorphic in* D_δ *and* D_ε *respectively* $(\delta > \varepsilon)$, *satisfying*

(7) $$\varkappa(t) = \hat{K}(t), \quad \text{for} \quad t \in R^d,$$

$$|\varkappa(t)| \leq \exp(-k(\text{Re } t)) \quad \text{for} \quad t \in D_\varepsilon,$$

and

$$\hat{H}(t)\eta(t) = 1 \quad \text{for} \quad t \in R^d$$

$$|\eta(t)| \leq \exp(h(\text{Re } t)) \quad \text{for} \quad t \in D_\delta,$$

where k *and* h *are sub-additive.*

If V *and* W *are positive sub-additive functions satisfying*

(8) $$\sup_x (V(x) - 2\pi\delta|x|) < \infty, \quad \sup_x (W(x) - 2\pi\varepsilon|x|) < \infty,$$

and if every $\phi \in L^\infty(R^d)$ *that satisfies*

$$\phi(x) = O(\exp(-V(x))), \quad x \to \infty,$$

and

$$K * \phi(x) = O(\exp(-W(x))), \quad x \to \infty,$$

also fulfills

(9) $$H * \phi(x) = O(1/Q(x)), \quad x \to \infty,$$

then

(10) $$Q(x) = O(1) \inf_{\Omega \in R^d_+} \left\{ \exp(V(x) + h(\Omega)) + \exp(W(x) + h(\Omega) - k(\Omega)) \right\}.$$

If we specialize some of the assumptions the following corollary is obtained.

COROLLARY 3.2.3 :

Let $H, K \in L(R)$. *Assume that their Fourier transforms* $\hat{H}$ *and* $\hat{K}$ *can be holomorphically extended to*

$$\{t \in \mathbb{C} : |\text{Im } t| \leq \delta\} \quad \text{viz.} \quad \{t \in C : |\text{Im } t| \leq \varepsilon\},$$

where $\varepsilon < \delta$, *and that there are positive numbers* $a, b, A, B, \alpha < \beta$ *such that*

(11) $$a \leq |\hat{H}(t)| \exp(\alpha|t|) \leq A \quad \text{for} \quad |\text{Im } t| \leq \delta$$

(12) $$b \leq |\hat{K}(t)| \exp(\beta|t|) \leq B \quad \text{for} \quad |\text{Im } t| \leq \varepsilon.$$

If $\phi \in L^\infty(R)$ *satisfies* $\phi(x) = O(\exp(-V(x))), \; x \to \infty,$ *and* $K * \phi(x) =$

$= O(\exp(-W(x)))$, $x \to \infty$, <u>with sub-additive</u> V <u>and</u> W <u>fulfilling</u> (5) <u>and</u> (8), <u>then</u>

(13) $$H * \phi(x) = O(1) \exp(-\frac{\alpha}{\beta} W(x) - (1-\frac{\alpha}{\beta})V(x)), \quad x \to \infty .$$

<u>This result is best possible in the sense that no function tending to zero can be substituted for</u> 1 <u>in</u> $O(1)$ <u>on the right side of</u> (13).

Recalling the general remainder problem mentioned in the introduction, we find that the theorems given above give solutions of problems of that type. We shall later see that certain refinements are necessary in order to obtain precise estimates in several interesting cases. We now turn to the proofs of the theorems and their corollary.

We define U and T by (3.1.2) and (3.1.3) taking $\omega = \pi$. Hence

$$\hat{U}(t) = \hat{H}(t) \int (1 - \gamma(u/\Omega)) \exp(-\pi(u-t)^2) \, du.$$

<u>Proof of theorem 3.2.1.</u>

Our assumptions in theorem 3.2.1 imply that we can change the line of integration by substituting $t \to t + i\delta^*$, where $|\delta^*| = \delta$, when computing $U(x) = \int \exp(2\pi i x t)\hat{U}(t) \, dt$. Hence

(14) $$U(x) = O(\exp(-2\pi\delta|x|)) \int \exp(-h(t)) \int (1-\gamma(u/\Omega) \exp(-\pi(u-t)^2) \, du \, dt .$$

The sub-additivity of h implies that

$$-h(t) \leq -h(u) + h(u-t) ,$$

so changing the order of integration in (14) we obtain

$$U(x) = O(\exp(-2\pi\delta|x|)) \int (1-\gamma(u/\Omega) \exp(-h(u)) \, du \int \exp(h(v) - \pi v^2) \, dv ,$$

where the last integral is bounded (we have put $v = u-t$). By aid of conditions (3), (5) and the inequality

$$-V(x-y) \leq V(y) - V(x)$$

we have

$$U * \phi(x) = O(1) \int |U(y)| \exp(-V(x-y)) \, dy =$$

$$= O(1) \exp(-V(x)) \int (1-\gamma(u/\Omega) \exp(-h(u)) du .$$

That is the first term on the right in the conclusion (6) of the theorem 3.2.1. We shall now show that $|T * \psi|$ can be estimated by the second term in (6) if

$$\hat{T}(t) = (\hat{H}(t)/\hat{K}(t)) \int \gamma(u/\Omega) \exp(-\pi(u-t)^2) \, du .$$

Changing the domain of integration as above we obtain

$$T(x) = \int \exp(2\pi itx)\hat{T}(t)\, dt$$
$$= O(\exp(-2\pi\varepsilon|x|)\int \exp(-h(t)+k(t))\, dt \int \gamma(u/\Omega)\exp(-\pi(u-t)^2)\, du \, .$$

Now

$$-h(t) \leq -h(u) + h(u-t)$$

and

$$k(t) \leq k(u) + k(t-u),$$

so that changing the order of integration we obtain

$$T(x) = O(\exp(-2\pi\varepsilon|x|))\int \gamma(u/\Omega)\exp(-h(u)+k(u))\, du \int \exp(h(v)+k(v)-\pi v^2)\, dv$$

where the last integral is bounded. If we now form

$$T*\psi(x) = \int T(y)\psi(x-y)\, du$$

we find by aid of (9) and the sub-additivity of W, that

$$T*\psi(x) = O(\exp(-W(x)))\int T(y)\exp W(y)\, dy =$$
$$= O(\exp(-W(x)))\int \gamma(u/\Omega)\exp(-h(u)+k(u))\, du$$

We have obtained the second term on the right of (6). Our computations show that $U \in L(R^d)$ and $T \in L(R^d)$ and hence $H*\phi(x) = U*\phi(x) + T*\psi(x)$. The number Ω is at our disposal and hence we can take the infimum and the first theorem is proved.

Proof of theorem 3.2.2.

To prove the second theorem we reformulate its assumptions concerning ϕ in the following way. Every function $\phi \in L^\infty(R^d)$ which has a finite norm

$$|||\phi|||_1 = ||\phi \exp V||_\infty + ||(\check{K}*\phi) \exp W||_\infty$$

also satisfies

$$|||\phi|||_2 = |||\phi|||_1 + ||Q(H*\phi)||_\infty < \infty \, .$$

Both the spaces $B_i = \{\phi \in L^\infty : |||\phi|||_i < \infty\}$, $i=1,2$, are easily seen to be Banach spaces. Theorem 3.2.2 says that they have the same elements and hence we have a set which is a Banach space under two norms, one of which is dominated by the other. According to a well-known consequence of the open mapping theorem it follows that there is a λ such that

$$|||\phi|||_2 \leq \lambda |||\phi|||_1 \, .$$

Introducing the expressions for the norms we find that there is a numbre $\mu > 0$, so that

$$(15) \qquad \mu \| Q(H * \phi) \|_{\infty} \leq ||| \phi |||_{1} \leq \| \phi \exp V \|_{\infty} + \| (K * \phi) \exp W \|_{\infty}$$

for all ϕ. We shall show that ϕ can be chosen in such a way that (15) can be rewritten in the form of the conclusion of our second theorem.

With a constant y we define ϕ by

$$(16) \qquad H * \phi(x) = \exp(2\pi i \Omega(x-y) - \pi (x-y)^2) ,$$

where $\Omega \in R^d$. First of all we observe that

$$(17) \qquad |Q(y)| = |Q(y) \, H * \phi(y)| \leq \| Q(H * \phi) \|_{\infty} .$$

Our assumptions imply that

$$\hat{\phi}(t) = \hat{H}(t)^{-1} \exp(-2\pi i t y - \pi (t-\Omega)^2)$$

and since $\hat{\phi} \in L(R^d)$ it follows that $\phi \in L^{\infty}(R^d)$. Hence it follows from (8) by change of integration paths that

$$(18) \qquad \phi(x) = \int \exp(2\pi i x t) \hat{\phi}(t) \, dt = O(\exp(-2\pi \delta (x-y)) \exp(h(\Omega))) ,$$

the calculations with the sub-additive function h being exactly the same as in the proof of the first theorem.

If we now consider

$$\| \phi \exp V \|_{\infty} = \sup_{x} | \phi(x) \exp V(x) | ,$$

we find by aid of (18) and the sub-additivity of V that

$$(19) \qquad \| \phi \exp V \|_{\infty} = O(\exp(V(y) + h(\Omega))) .$$

In a similar way we get by aid of (7) that

$$K * \phi(x) = \int \exp(2\pi i x t) K(t) \phi(t) \, dt = O(\exp(-2\pi \varepsilon \, |x-y|) \exp(-k(\Omega) + h(\Omega))$$

and

$$(20) \qquad \| (K * \phi) \exp W \|_{\infty} = O(1) \exp(W(y) - k(\Omega) + h(\Omega)) .$$

Insertion of the estimates (19) and (20) in (15) immediately gives the conclusion (14) of our theorem, since $|Q(y)| \leq \|Q(H * \phi)\|_\infty$ by (17).

Proof of corollary 3.2.3.

To obtain the corollary we substitute $\alpha|t|$ for $h(t)$ and $\beta|t|$ for $k(t)$ in our theorems taking $d=1$. Nothing prevents us from taking γ equal to the characteristic function of the interval $[-1, 1]$.

Since

$$\int_{|t| \geq \Omega} \exp(-\alpha|t|)\, dt = O(\exp(-\alpha\Omega))$$

and

$$\int_{|t| \leq \Omega} \exp(\beta|t| - \alpha|t|)\, dt = O(\exp(\beta-\alpha)\Omega) ,$$

we infer from (6) by taking

$$\beta\Omega = W(x) - V(x)$$

that (13) holds. The same choice of Ω in (10) gives

$$Q(x) = O(1) \exp((1-\frac{\alpha}{\beta})V(x) + \frac{\alpha}{\beta} W(x)) ,$$

and the corollary is proved.

3.3 Application of the results to theorem 2.1

It is well-known that the assumptions of theorem 2.1 also imply estimates for the Riesz means of the function. With the notations of that theorem we can prove that

(1) $$R_p * \phi(x) = O(W(x)^{-1-p}) ,$$

where $R_p = K_3$ is the kernel given in 1.5 and characterized by

$$\hat{R}_p(t) = \Gamma(1 + 2\pi it)/\Gamma(1 + p + 2\pi it) .$$

For simplicity we only consider the case when p is a positive integer so that $\hat{R}_p(t)^{-1}$ is a polynomial of degree p.

We are going to show that (1) can be derived by the methods of the preceding sections. We first observe that the sub-additivity of W, implies the sub-additivity of

$$V(x) = \log(1 + W(x)) ,$$

since

$$V(x + y) = \log(1 + W(x+y)) \leq \log(1 + W(x)) + \log(1 + W(y)) .$$

Hence we know that

$$\phi(x) = O(\exp(-V(x))) ,$$

and that

$$\psi(x) = K * \phi(x) = O(\exp(-\exp V(x))) ,$$

where $\hat{K}(t) = \Gamma(1 + 2\pi it)$.

We apply the formulas (3.1.1-3), i.e.

$$R_p * \phi = U * \phi + T * \psi ,$$

with

(2) $$\hat{U}(t) = \hat{R}_p(t) \int \exp(-\pi(t-u)^2)(1-\gamma(u/\Omega))\, du$$

and

$$\hat{T}(t) = (\hat{R}_p(t)/\hat{K}(t)) \int \exp(-\pi(t-u)^2)\, \gamma(u/\Omega)\, du .$$

Since $\hat{R}_p(t)/\hat{K}(t) = \Gamma(1 + p + 2\pi it)^{-1}$ which has a majorant $C \exp(d|t|)$, it follows from our considerations in section 3.2 that

(3) $$T*\psi(x) = O(\exp(d\Omega - \exp V(x))) .$$

If we can show that

$$U * \phi(x) = O(\Omega^{-p} \exp(-V(x))) ,$$

the choice $d\Omega = \frac{1}{2} \exp V(x)$ gives us

$$R_p * \phi(x) = U * \phi(x) + T*\psi(x) = O(\exp(-(p+1)\, V(x))) ,$$

and (1) is proved.

Since $V(x) = \log(1 + W(x))$ does not increase as rapidly as x, and $\hat{U}$ is holomorphic in a strip around the real axis, we infer as in the previous section that

$$U * \phi(x) = O(\| U \exp 2\pi|\sigma\cdot| \| \exp(-V(x))) ,$$

and it remains to prove that $\hat{U}(t \pm i\sigma)$ are Fouriertransforms of functions with L_1-norm of the order Ω^{-p}.

That can be done by aid of Carlson-Beurling's inequality

(4) $$\| f \|_1^2 \leq c \| \hat{f} \|_2 \; \| \hat{f}' \|_2 .$$

The integral in formula (2) is bounded and it is of the order $\exp(-k\Omega^2)$, if $|t| \leq \frac{1}{2}\Omega$. Let us restrict our attention to the case $0 < \sigma < \delta(2\pi)^{-1}$. Then

$$\| \hat{U}(\,.\, + i\sigma) \|_2^2 \leq O(1) \int_{|t| \geq \Omega/2} |\hat{R}_p(t+i\sigma)|^2 \, dt + \Omega \exp(-k\Omega^2) =$$

$$= O(\Omega^{-2p+1}) ,$$

since $\hat{R}_p^{-1}$ is a polynomial of degree p. We compute $\hat{U}'$ and get

$$\hat{U}'(t+i\sigma) = \hat{R}_p'(t+i\sigma) \int \exp(-\pi(t+i\sigma-u)^2)(1-\gamma(u/\Omega))\, du +$$

$$+\Omega^{-1}\hat{R}_p(t+i\sigma) \int_{\Omega \leq |u| \leq 2\Omega} \gamma'(u/\Omega) \exp(-\pi(t+i\sigma-u)^2)\, du.$$

Applying Minkowski's inequality and the same estimates as above we get

$$\| \hat{U}'(. + i\sigma) \|_2^2 = O(\Omega^{-2p-1}) ,$$

since *differentiation of* $\hat{R}_p$ increases the rate of decrease by one.

Introducing the two estimates in (4) we find that

(5) $$\| U \exp 2\pi|\sigma .| \| = O(\Omega^{-p})$$

and formula (1) is proved.

On the precision of theorem 2.1

The estimates we have applied, can be used to obtain an alternative proof of theorem 2.1 , if we strengthen the tauberian condition (2.1.2) to

(6) $$|\phi'(x)| = O(1).$$

We use the following device which cannot immediately be extended to problems in several dimensions. We define $\chi \in L^\infty(R)$ by $\chi = \phi + \phi'$. It is easy to see that there is an $H \in L$ with

$$\hat{H}(t) = (1 + 2\pi it)^{-1}$$

such that $\phi = H * \chi$. In fact $H = R_p$ for $p = 1$.

The problem of theorem 2.1 is now transformed into a new one : to estimate $H*\chi(x)$, knowing that $\chi(x) = O(1)$ and

$$\psi(x) = (K * H) * \chi(x) = O(\exp -W(x)) .$$

That is a problem of the type considered in the previous section, and as above we write

$$H * \chi = U * \chi + T * \psi ,$$

where

$$\hat{U}(t) = (1+2\pi it)^{-1} \int (1-\gamma(u/\Omega)) \exp(-\pi(u-t)^2)\, du$$

and

$$\hat{T}(t) = \hat{K}(t)^{-1} \int \gamma(u/\Omega) \exp(-\pi(u-t)^2)\, du .$$

Under the same assumptions but with slightly different notation we obtained (3). In the present case the corresponding estimate is

$$T*\psi(x) = O(1)\exp(d\Omega - W(x)) \ ,$$

and (5) shows that

$$||U|| = O(\Omega^{-1}) \ .$$

Hence

$$H*\chi(x) = O(1)\{\Omega^{-1} + \exp(d\Omega - W(x))\}$$

and $d\Omega = \frac{1}{2}W(x)$ gives

$$H*\chi(x) = O(1/W(x)) \ .$$

As we have indicated earlier theorem 3.2.2 shows that the result of theorem 2.1 is best possible. We shall in fact prove the slightly stronger result that the conclusion cannot be improved, if we assume the tauberian condition (6).

We consider the bound

$$H*\chi(x) = O(1/Q(x)) \ ,$$

if $(K*H)*\chi(x) = O(\exp(-W(x)))$, where $\hat{K}(t) = \Gamma(1 + 2\pi it) = O(\exp(-b|t|))$ and $H(t) = (1 + 2\pi it)^{-1}$.

In the conclusion (3.2.10) of theorem 3.2.2 we introduce

$$h(u) = \log(c + 2\pi|u|)$$

and $h(u)-k(u) = -b|u|$. Since $V=0$ formula (3.2.10) reads

$$Q(x) = O(1)\inf_{\Omega}\{c + 2\pi\Omega + \exp(W(x)-b\Omega)\}$$

and taking $\Omega = (2/b)W(x)$ we get

$$Q(x) = O(W(x)) \ .$$

Hence the estimate given in theorem 2.1 cannot be improved and the same method works also for the Riesz means.

A direct construction showing the precision in the case of the Laplace transform is given by Korevaar 1951.

CHAPTER 4

SOME PRECISE THEOREMS IN SEVERAL DIMENSIONS

4.1 Introduction

In the previous chapters we obtained some remainder estimates. The first one is a precise general remainder theorem corresponding to Hardy-Littlewood's theorem with the tauberian condition in the most general form (due to Schmidt 1925 in the classical case). A closer inspection of theorems 3.2.1 and 3.2.2 reveals that a direct application of these general results will not produce sharp results in many cases of specific interest. As we shall see in this chapter that does not depend on defects of the method, but is very much in the heart of the matter. To obtain best possible estimates we have to invoke stronger assumptions on the kernels.

From the discussion in section 3.3 it should be clear that the weak tauberian conditions make complicated methods necessary. Having shown how to handle such problems in the proof in chapter 2 of the remainder analogue of Hardy-Littlewood's theorem, I don't find it worth while to formulate the theorems of this chapter in the more general way. I hence restrict my attention to theorems, where the tauberian condition takes the form

$$\| \operatorname{grad} \phi \| \leq 1 .$$

We are now ready to state our first theorem.

4.2 A general remainder theorem with the tauberian condition $\|\operatorname{grad} \phi\| \leq 1$.

We apply the notation introduced in section 3.1. Hence $x \to \infty$ means that $x_k \to \infty$ for all $k = 1, 2, \ldots, d$ and if $\delta \in R_+^d$, then $D_\delta = \{ z \in ¢^d : |\operatorname{Im} z| < \delta \}$.

THEOREM 4.2.1 :

Let $K \in L(R^d)$ and let $\phi : R^d \to ¢$ be bounded and continously differentiable with

(1) $$\|\operatorname{grad} \phi\| \leq 1 .$$

Suppose that

(2) $$\psi(x) = K * \phi(x) = O(\exp(-W(x))) , \quad x \to \infty ,$$

where W is positive, sub-additive and fulfills $\int \exp(W(x) - 2\pi\beta|x|)\, dx < \infty$ for some $\beta \in R_+^d$. Let $\delta \in R_+^d$ satisfy $\beta < \delta$.

I. *If there is a* $\varkappa$ *holomorphic in* D_δ *such that*

(3) $$\hat{K}(t)\varkappa(t) = 1 \quad \underline{\text{for}} \quad t \in R^d$$

and

(4) $$|\varkappa(t)| \leq g(||\mathrm{Re}\ t||) \quad \underline{\text{for}} \quad t \in D_\delta ,$$

where $g : R_+ \to R_+$ *is non-decreasing, then as* $x \to \infty$ *it holds that*

(5) $$\phi(x) = O(1) \inf_{\Omega>0} \left\{\Omega^{-1} + \Omega^{d/2} \int g(||v|| + \Omega)\exp(-\pi v^2)\ dv\ \exp(-W(x))\right\} .$$

II. *Conversely, if there is a* λ *holomorphic in* D_δ *and such that*

(6) $$\hat{K}(t) = \lambda(t) \quad \underline{\text{for}} \quad t \in R^d ,$$

and

(7) $$|\lambda(t)| \leq h(||\mathrm{Re}\ t||) \quad \underline{\text{for}} \quad t \in D_\delta$$

with $h : R_+ \to R_+$, *non-increasing, and if every function* $\phi \in L^\infty(R^d)$ *with* $||\mathrm{grad}\ \phi|| \leq 1$ *and* $K * \phi(x) = O(\exp(-W(x)))$, *also satisfies* $\phi(x) = O(1/S(x))$, $x \to \infty$, *then as* $x \to \infty$, *it holds that*

(8) $$S(x) = O(1) \inf_{\Omega>0} \left\{\Omega + \int h(||v|| - \Omega)\ \exp(-\pi v^2)\ dv\ \exp(W(x))\right\} .$$

The starting point of the proof is an appropriate generalization to the d-dimensional case of the approach in section 2.1.

The important condition on χ in the derivation of (2.2.1) and (2.2.2) is that $\hat{\chi}$ has compact support. For our present purposes we replace $\hat{\chi}$ by $\gamma(\Omega^{-1}(.))$ where the C^∞-function $\gamma : R^d \to R$ has support in $\{x : ||x|| \leq 1\}$ and positive Fouriertransform. In exactly the same way as (2.2.1) and (2.2.2) were proved, we now obtain that for all $\eta \in R^d$ it holds that

(9) $$\int \phi(x-y)\ E(y)\ \hat{\gamma}(\Omega(\eta-y))\,\Omega\ dy = Q * \psi(x) ,$$

where

(10) $$\hat{Q}(t) = \varkappa(t) \int \hat{E}(t-u)\ \gamma(\Omega^{-1}u)\ e^{-2\pi i u\eta}\ du .$$

The functions ϕ, ψ and $\varkappa$ are given in theorem 4.2.1 and $E(y) = \exp(-\pi y^2)$. Proceeding as in section 2.3 we introduce

$$M_x = \sup_y |\phi(x-y)\ E(y)| \geq |\phi(x)| ,$$

and take η so that $|\phi(x-\eta)\ E(\eta)| \geq \frac{3}{4} M_x$. We normalize by choosing

$$\gamma(0) = \int \hat{\gamma}(x)\ dx = 1 ,$$

and pick a number R so that $\int_{\|x\| \leq R} \hat{\gamma}(x)\ dx > \frac{2}{3}$. Similarly as (2.3.1) follows from (2.2.1), it now follows from (9), that

$$(11) \qquad |\phi(x)| \leq 4 \sup_{\|v-y\| \leq R\Omega^{-1}} |\phi(x-y)E(y) - \phi(x-v)E(v)| + 6|Q*\psi(x)| ,$$

Since $\|\text{grad}\ \phi\|$, $\|\text{grad}\ E\|$, $|\ \phi\ |$, and $|\ E\ |$ are all bounded we have proved that

$$(12) \qquad |\phi(x)| \leq O(1) \inf_{\Omega} \{\Omega^{-1} + |Q_{\Omega}*\psi(x)|\} ,$$

where $Q_{\Omega} = Q$ is given by (10).

The next step is to estimate $Q*\psi$. As in the earlier proof we change the paths of integration with respect to t in order to show the exponential decrease in x of Q. We split the integral with respect to t in (10) in the sum of integrals over the generalized octants $\{Z_m\}_1^{2^d}$, where for any m, the assumption $x, y \in Z_m$ implies that the components x_k and y_k have the same sign, $k = 1, 2, \ldots, d$. Let us call the integrals $\{I_m\}$ and write

$$Q(x,\Omega) = \sum_{m=1}^{2^d} \int_{Z_m} \varkappa(t)\exp(2\pi ixt) \int \exp(-2\pi i\eta u - \pi(u-t)^2)\ \gamma(u/\Omega)\ du\ dt .$$

To estimate I_m we substitute $t \to t+i\beta^{(m)}$, where $\beta^{(m)} \in D_\delta$ so that

$$I_m(x) =$$

$$= \exp(-2\pi\sum \beta_k|x_k|)\int \varkappa(t+i\beta^{(m)})\exp(2\pi ixt)\int \exp(-2\pi i\eta u - \pi(u-t-i\beta^{(m)})^2\gamma(u/\Omega)\ du\ dt .$$

We now form $I_m*\psi(x)$ after having transformed the integral with respect to u by the transformation $u=t-v$. The change of order of integration is legitimate by Fubini's theorem under our assumptions. We get

$$(13) \qquad |I_m*\psi(x)| = |\int I_m(y)\psi(x-y)\ dy| \leq$$

$$\leq \int \exp(-\pi v^2) \int |\psi(x-y)\exp(-2\pi\beta|y|) \int \exp(2\pi i(y-\eta)t)\gamma((t-v)/\Omega)\varkappa(t+i\beta^{(m)})\ dt|\ dy\ dv .$$

We apply the Cauchy-Bunyakovski inequality to the integral with respect to y and get

(14) $$I_m * \Psi(x) = O(1) \int \exp(-\pi v^2) \| R(.,x) \|_2 \| \hat{T}(.,v,\Omega) \|_2 dv ,$$

where

$$R(y,x) = |\Psi(x-y)| \exp(-2\pi\beta|y|)$$

and

$$T(t,v,\Omega) = \gamma((t-v)/\Omega) \varkappa(t+i\beta^{(m)}) .$$

Since

$$\Psi(x-y) = O(\exp(-W(x-y))) = O(\exp(-W(x) \exp(W(y))),$$

our assumptions on β and W imply that $\| R(.,x) \|_2 = O(\exp(-W(x)))$. Parseval's formula gives after insertion of estimate in (4) that

$$\| \hat{T}(.,v,\Omega) \|_2^2 = \int \gamma((t-v)/\Omega)^2 g(\| t \|)^2 \, dt \leq \Omega^d g(\| v \| + \Omega \sqrt{d})^2 .$$

Introducing these estimates in (14) we get

$$Q * \Psi(x) = \sum_1^{2^d} I_m * \Psi(x) = O(\Omega^{d/2}) \int g(\| v \| + \Omega \sqrt{d}) \exp(-\pi v^2) \, dv \exp(-W(x)) .$$

A trivial transformation $\Omega\sqrt{d} \to \Omega$ gives the conclusion (5) of the first part of our theorem.

The second part of the theorem is proved similarly as the corresponding parts of the theorem in chapter 3.

We consider the Banach space B of all continuosly differentiable functions ϕ with a finite norm

$$|||\phi|||_1 = \| \phi \|_\infty + \| (\| \text{grad } \phi \|) \|_\infty + \| (K * \phi) \exp W \|_\infty .$$

The assumptions of the second part of our theorem imply that every element of B also has a finite norm

$$||| \phi |||_2 = ||| \phi |||_1 + \| S \phi \|_\infty ,$$

and hence according to the well-known result for Banach spaces, there is a constant m such that

$$||| \phi |||_2 \leq m ||| \phi |||_1 ,$$

that is

$$\| S \phi \|_\infty \leq (m-1) ||| \phi |||_1 .$$

This inequality written in the form

$$(15)\qquad |S(\omega)| \leq O(1)\ \inf \left\{ \|\phi\|_\infty + \|(\|\operatorname{grad}\phi\|)\|_\infty + \|(K*\phi)\exp W\|_\infty \right\} ,$$

where the lower bound is taken over all admissible ϕ with $|\phi(\omega)| = 1$, will be the starting point for the proof of (8).

We choose

$$\phi(x) = \exp(2\pi i\sigma(x-\omega) - \pi(x-\omega)^2) ,$$

so that evidently $\|\phi\|_\infty = 1$. The value of $\sigma \in R^d$ will be chosen later. Differentiation gives

$$D_j\phi(x) = 2\pi i\,\sigma_j\,\phi(x) - 2\pi(x_j - \omega_j)\,\phi(x)$$

and hence

$$\|D_j\,\phi\| \leq O(|\sigma_j| + 1) .$$

If we take $|\sigma_j| \leq V$ for all j then

$$(16)\qquad \|\operatorname{grad}\phi\| \leq O(1)(1+V) .$$

It remains to estimate the last term on the right in (15). We note that

$$\hat{\phi}(t) = \exp(-2\pi it\omega - \pi(t-\sigma)^2) ,$$

so that

$$K*\phi(x) = \int \exp(2\pi itx)\hat{\phi}(t)\,\lambda(t)\,dt =$$
$$= \int \exp(2\pi it(x-\omega) - \pi(t-\sigma)^2)\,\lambda(t)\,dt =$$
$$= \exp(-2\pi\beta|x-\omega|)\sum_{m=1}^{2^d} \int_{Z_m} \exp(2\pi it(x-\omega) - \pi(t+i\beta^{(m)} - \sigma)^2)\lambda(t+i\beta^{(m)})\,dt$$

On account of the sub-additivity of W we have

$$W(x) \leq W(x-\omega) + W(\omega)$$

and hence our assumptions on W and β give

$$(17)\qquad \|(K*\phi)\exp W\|_\infty \leq \exp W(\omega)\int h(\|t\|)\exp(-\pi(t-\sigma)^2)\,dt .$$

Recalling that $|\sigma_j| \leq V$, we get by insertion of (16) and (17) in (15) that

$$|Q(\omega)| = O(1)\left\{1 + V + \int h(\|t\| - \sqrt{d}\,V)\exp(-\pi t^2)\,dt\,\exp W(\omega)\right\}$$

and the substitution $\sqrt{d}\,V = \Omega$ gives the conclusion (8) of the second part of the theorem.

We shall now see that theorem 4.2.1 implies a number of special results obtained by other authors.

EXAMPLE 1 : If $g(r) = C\exp(cr)$, the first part of the theorem gives

$$|\phi(x)| = O(1)\inf\left\{\Omega^{-1} + \Omega^{d/2}\exp(c\Omega - W(x))\right\} .$$

Taking $c\Omega = \frac{1}{2}W(x)$, we get

$$|\phi(x)| = O(1/W(x)) , \quad x \to \infty .$$

The second part of the theorem shows that this result is the best possible, if

$$|\hat{K}(t)| \leq C\exp(-c\|t\|) \quad \text{for} \quad t \in D_\delta .$$

This example applies to the most studied case of Laplace and Stieltjes transforms, if the remainders do not decrease more rapidly than exponentials. For one-dimensional results of this type see e.g. Freud 1951, Korevaar 1951, 1954, Subhankulov 1960. The d-dimensional Laplace transform is treated in Frennemo 1966, 1967. Some other kernels with this behavior can be found in the list at the end of the first chapter.

EXAMPLE 2 : If $g(r) = C\exp(cr^2)$, $c < \pi$, the first part of the theorem gives

$$|\phi(x)| = O(1)\inf\left\{\Omega^{-1} + \exp(\lambda(c)\Omega^2 - W(x))\right\}$$

and hence

$$|\phi(x)| = O(W(x)^{-1/2}) .$$

The second part shows that this result is best possible, if

$$|\hat{K}(t)| \leq C\exp(-c\|t\|^2) \quad \text{for} \quad t \in D_\delta .$$

That is true for instance in the case of the Weierstrass transform. The restriction $c < \pi$ is immaterial and can be removed by a change of variables.

More intriguing cases arise, if we turn to kernels with majorants of polynomial growth, since the differences between formulas (5) and (8) become of importance.

EXAMPLE 3 : If $g(r) = C(1+r)^m$, $m > 0$, we get

$$|\phi(x)| = O(1) \inf \{\Omega^{-1} + \Omega^{d/2} \int (1+\Omega+\|v\|+\|\beta\|)^m \exp(-\pi v^2)\, dv \exp(-W(x))\},$$

hence

$$|\phi(x)| = O(1) \inf \{\Omega^{-1} + \Omega^{m+d/2} \exp(-W(x))\} .$$

Choosing $V = (\exp(-W(x))/(m+\frac{d}{2}+1))$ we obtain

(18) $$\phi(x) = O(\exp(-W(x)/(m+\frac{d}{2}+1))) .$$

The corresponding one-dimensional result is the first general remainder theorem given by Beurling 1938.

Application of the second part of the theorem gives that the best possible bound under our assumptions can certainly not be better than

(19) $$\phi(x) = O(\exp(-W(x)/(m+1))) .$$

We are next going to show that the gap between the two estimates (18) and (19) does not depend on defects in our methods. In fact it turns out that both the results are in a sense best possible, so that both bounds are of interest for different kernels in the class considered.

We shall give some results, which justify the above statements.

4.3 More precise results when $\hat{K}(t)^{-1}$ is of polynomial type

We shall first prove that some additional assumptions, which are fulfilled in many special cases, make it possible to sharpen the results of theorem 4.2.1. We also show that the restriction to sub-additive W can be removed, and that arbitrarily rapidly decreasing remainders can be dealt with, if we assume that the conditions on the kernel hold in the whole plane.

THEOREM 4.3.1 :

Let the assupmtions of theorem 4.2.1 be satisfied. Assume further that there is an $n \geq 1 + [d/4]$ such that

(1) $$|D^\alpha \kappa(t)| \leq C(1+\|t\|)^{n-|\alpha|} \quad \text{for } t \in D_\delta$$

<u>and every differential operator</u> D^α <u>of order</u> $|\alpha| \leq 1 + [d/4]$. <u>Then it follows that</u>

(2) $$\phi(x) = O(\exp(-W(x)/(n+1))) \ , \quad x \to \infty \ .$$

It might be remarked that it is not necessary to assume that all these derivatives are bounded, since they are dependent. I have tried to give a simple formulation, which is applicable to the examples given below.

In the proof we need the following lemma.

<u>LEMMA 4.3.2 :</u>

<u>If the function</u> $F \in L^2(R^d)$ <u>and</u> $\Delta^s F \in L^2(R^d)$, <u>where</u> Δ <u>is the Laplacian and</u> $s = 1 + [d/4]$, <u>then the Fourier transform</u> $\hat{F}$ <u>of</u> F <u>belongs to</u> $L(R^d)$ <u>and</u>

(3) $$\|\hat{F}\|_1 \leq C_d \|F\|_2^{1-d/(4s)} \|\Delta^s F\|_2^{d/(4s)} \ .$$

If the right hand side is finite, then F has a Fourier transform $\hat{F}$ in L^2 such that $\|x\|^{2s}\hat{F}(x)$ belongs to L^2. If B_ω is a ball with radius ω, then

$$\|\hat{F}\|_1 = \int_{R^d} |\hat{F}(x)|\, dx = \int_{\|x\|\leq\omega} + \int_{\|x\|\geq\omega} |\hat{F}|\, dx \leq$$

$$\leq \Big(\int_{B_\omega} dx \int_{B_\omega} |\hat{F}(x)|^2\, dx\Big)^{1/2} + \Big(\int_{\complement B_\omega} \|x\|^{-4s} dx \int_{\complement B_\omega} \|x\|^{4s} |\hat{F}(x)|^2 dx\Big)^{1/2}$$

$$\leq A_d\, \omega^{d/2} \|F\|_2 + C_d\, \omega^{d/2-2s} \|\Delta^s F\|_2 \ ,$$

by Cauchy-Bunyakovski's inequality. Taking $\omega = (\|\Delta^s F\|_2 / \|F\|_2)^{1/(2s)}$ we obtain (3).

To prove theorem 4.3.1 we now proceed in exactly the same way as in the proof of theorem 4.2.1 up to formula (4.2.13). Instead of that formula we write more explicitly

(4) $$I_m * \psi(x) = \qquad (dw = dt\ du\ dv)$$

$$= \int \exp(2\pi i \eta v - \pi(v+i\beta^{(m)})^2) \int_{Z_m} \psi(x-y) \exp(-2\pi\beta|y|) \int \exp(2\pi i t(y-\eta)) \varkappa(t+i\beta^{(m)}) \gamma(\tfrac{t-v}{\Omega})\, dw.$$

To our previous assumptions on γ we add $0 \leq \gamma \leq 1$. We now apply our lemma to get an estimate of the L_1-norm of the inner integral as a function of y. Taking

$$F(t) = \gamma((t-v)/\Omega)\, \varkappa(t+i\beta^{(m)}) \ ,$$

we have to estimate $||F||_2$ and $||\Delta^s F||_2$. Since $|\gamma|$ is bounded by 1, we get

(5) $$||F||_2^2 \le \int_{||t-v|| \le \Omega} |\varkappa(t+i\beta^{(m)})|^2 dt \le O(1)(1+\Omega+||v||+||\beta||)^{2n}\Omega^d .$$

Next we consider

$$\Delta^s F(t) = \sum_{|p|+|q|=2s} D_t^p \gamma((t-v)/\Omega) D_t^q \varkappa(t+i\beta^{(m)}) .$$

By Minkowski's inequality and our assumptions we obtain

(6) $$||\Delta^s F||_2 \le \sum || D_t{}^p \gamma((t-v)/\Omega) D_t{}^q \varkappa(t+i\beta^{(m)}) ||_2 \le$$
$$\le O(1) \sum \Omega^{-|p|}(1+||v||+\Omega+||\beta||)^{n-|q|+\frac{d}{2}} \le$$
$$\le O(1) \sum \Omega^{-2s}(1+\Omega+||v||+||\beta||)^{n+\frac{d}{2}} .$$

Inserting the estimates (5) and (6) in (3) we get

$$||\gamma((.-v)/\Omega)\varkappa(.+i\beta^{(m)})||_1 \le O(1)\Omega^{-d^2/(8s)}(1+\Omega+||v||+||\beta||^{\,n+d^2/(8s)}) .$$

Returning to formula (4) and observing that as before

$$\int |\Psi(x-y)\exp(-2\pi\beta|y|)|dy = O(1)\exp(-W(x)) ,$$

we find that

$$I_m * \Psi(x) =$$
$$= O(1)\exp(-W(x))\Omega^{-d^2/(8s)} \int (1+\Omega+||v||+||\beta||)^{n+d^2/(8s)} \exp(-\pi v^2) dv .$$

Adding these inequalities for different m we get the same estimate for $Q*\Psi(x)$ and formula (4.2.12) gives

$$\phi(x) = O(1) \inf\{\Omega^{-1}+\Omega^n \exp(-W(x))\} .$$

Taking $\Omega = \exp(W(x)/(n+1))$ we obtain the conclusion (2) of our theorem.

As we found in the discussion of example 3 in the preceding section, the second part of theorem 4.2.1 shows that this result is the best possible.

EXAMPLE 1 : If $K(x) = \exp(-c||x||)$, $x \in R^d$, then

$$\hat{K}(t) = \Gamma(\tfrac{1}{2}(d+1))\pi^{-1/2}c(c^2+4\pi^2t^2)^{-\frac{1}{2}(d+1)}$$

(cf. Bochner 1949, p.69, Erdelyi et al. 1954, p.182). Since

$$\hat{K}(t)^{-1} = O((1 + \|t\|)^{d+1}$$

and

$$D^{\alpha}(\hat{K}(t)^{-1}) = O((1 + \|t\|)^{d+1-|\alpha|})$$

for any differential operator of order $|\alpha|$, theorem 4.3.1 gives the precise estimate

$$\phi(x) = O(1)\ \exp(-W(x)/(d+2)) \ .$$

EXAMPLE 2 : An important example in one dimension is the convolution kernel corresponding to Cesàro and Riesz means. With $p>0$ and $\mathrm{Re}\ s>0$, we have

$$K(x) = \begin{cases} \Gamma(s)^{-1}\exp(-pt)(1-\exp(-x))^{s-1} & \text{for } x > 0 \\ 0 & \text{for } x \leq 0 \end{cases}$$

and

$$\hat{K}(t) = \Gamma(p+s+2\pi it)/\Gamma(p+2\pi it) = O(1)(1+|t|)^{\mathrm{Re}\ s} \ .$$

Also in this case theorem 4.3.1 is applicable and we get estimates of the form

$$\phi(x) = O(1)\ \exp(-W(x)/(1+\mathrm{Re}\ s)) \ .$$

This result implies known convexity theorems for the Riesz means (Chandrasekharan-Minakshisundaram 1952, p. 19 ff).

An one-dimensional example, to which theorem 4.3.1 does not apply is given as example 7 in section 1.5.

Theorem 4.3.1 shows that if the Fourier-transform of a convolution kernel is in a certain sense regular, then the best possible bound (4.2.19) is precise. One might ask if the bound (4.2.18) given in connection with example 4.2.3 is relevant for some class of kernels. I cannot give a complete answer to this question, but I conclude this chapter by showing that there is a class of kernels for which this bound cannot be substantially improved (cf. Frennemo 1967 b).

THEOREM 4.3.3 :

In the assumptions of theorem 4.2.1 we change the second part as follows.

II. There is a λ holomorphic in D_δ and such that

(6) $$\hat{K}(t) = \lambda(t) \quad \underline{\text{for}} \quad t \in R^d$$

<u>and</u>

(7') $$\lambda(t)\exp(2\pi i t\eta) = D_* P(t,\eta)\ , \qquad t \in D_\delta$$

<u>where</u> $D_* = \dfrac{\partial^d}{\partial t_1, \ldots, \partial t_d}$ <u>and</u>

(7") $$|P(t,\eta)| \leq C(1+\|t\|)^{-m}\ , \qquad t \in D_\delta\ , \qquad 0 < m\ ,$$

<u>with</u> C <u>independent of</u> t <u>and</u> η. <u>If every function</u> $\phi \in L^\infty(R^d)$ <u>with</u> $\|\text{grad}\ \phi\| \leq 1$ <u>and</u> $K * \phi(x) = O(\exp(-W(x)))$ <u>also satisfies</u> $\phi(x) = O(1/S(x))$, $x \to \infty$, <u>then as</u> $x \to \infty$, <u>it holds that</u>

$$S(x) = O(\exp(W(x)(m+1+d-\varepsilon)^{-1}))$$

<u>for every</u> $\varepsilon > 0$.

For the proof we apply formula (4.2.15) with

$$\phi(x) = \exp(2\pi i\sigma(x-\omega) - \tfrac{1}{2}\rho^2(x-\omega)^2)\ , \qquad \rho > 1$$

so that

$$\|\phi\|_\infty = 1, \qquad \|(\|\text{grad}\ \phi\|)\|_\infty = O(V+\rho)$$

if $|\sigma_j| \leq V$. Since

$$\hat{\phi}(t) = \rho^{-d}\exp(-2\pi i t\omega - \pi\rho^{-2}(t-\sigma)^2)$$

we find in a similar way as in the previous proof that

$$K * \phi(x) =$$

$$= \exp(-\beta|x-\omega|)\int \exp(2\pi i t(x-\omega))\eta(t+i\beta)\exp(-\pi\rho^{-2}(t+i\beta-\sigma)^2)\frac{dt}{\rho^d} =$$

$$= \rho^{-d}\exp(-\beta|x-\omega|)\int P(t+i\beta)D_*\exp(-\pi\rho^{-2}(t+i\beta-\sigma)^2)\ dt =$$

$$= O(1)\exp(-\beta|x-\omega|)\Big((1+\|\sigma\|)^{-m}\rho^{-d} + \int_{\|t\| \geq \pi\|\sigma\|} \|t\|^{2d}\exp(-\pi\rho^{-2}\|t\|^2)\ dt\ \rho^{-3d}\Big).$$

We take $\rho = \|\sigma\|^{1+\varepsilon/d}$ and get

$$K * \phi(x) = O(1)\exp(-\beta|x-\omega|)\ \|\sigma\|^{-m-d+\varepsilon}\ .$$

Inserting the estimates in (4.2.15) we find after use of the sub-additivity of W in the usual way that

$$|S(\omega)| \leq C\left\{V + V^{-m-d+\varepsilon} \exp W(\omega)\right\},$$

and the theorem follows by taking

$$V = \exp(W(\omega)(m+1+d-\varepsilon)^{-1}) .$$

To see that this shows that the bound (4.2.18) is of the proper order of magnitude we restrict our attention to an one-dimensional example.

EXAMPLE 3 : If

$$K(x) = \exp(-\beta x) \sin(\exp x) , \quad 0<\beta<1 ,$$

then

$$\hat{K}(t) = -\Gamma(-\beta-2\pi it) \sin \tfrac{\pi}{2}(\beta+2\pi it) = O(|t|^{-\frac{1}{2}-\beta})$$

for real β .

By aid of Stirling's formula it can be seen that

$$\int_{\tau}^{\infty} \exp(2\pi it\eta)\hat{K}(t)\, dt = O(\tau^{-\beta+\varepsilon})$$

for large τ independently of η. The crucial step is the observation that

$$\left|\int_{\tau}^{\infty} \frac{\cos(2\pi t \log t - \eta t)}{t^{\gamma}}\, dt\right| \leq C\tau^{\frac{1}{2}-\gamma} .$$

If (4.2.18) and theorem 4.3.3 are applied to this example we evidently get an example of the type desired.

CHAPTER 5

RAPIDLY DECREASING REMAINDERS

5.1 Introduction

Up to now we have only considered remainder theorems in which $\Psi(x) = K*\phi(x)$ decreases more slowly than $\exp(-ax)$ for some positive a. To obtain sharp estimates in the case of remainders of a more rapid decrease we have to invoke conditions on $\hat{K}(t)^{-1}$ in the whole upper half-plane. Hence the interesting results deal with more restricted classes of kernels than before. The original Wiener theorem holds for a large set of kernels, but we have seen in chapters 3 and 4 that the precise remainder estimates depend on the growth of $\hat{K}(t)^{-1}$ in a certain neighborhood of the reals. The restrictions we have to impose when dealing with very rapidly decreasing remainders do not leave much more than one kernel of interest in the set, so we are gradually passing over from general to special tauberian theorems.

Accepting that factthe purpose of this chapter is to show that the methods developed in previous chapters for general theorems, can easily be adopted to the case of rapidly decreasing remainders.

For every kernel there is a certain critical order of decrease of the remainder. If the remainder goes to 0 more rapidly, then it has to be identically 0. This phenomenon has been observed by several authors and will be considered in next chapter, where we give some general results.

In this chapter we only give two theorems, the first being applicable to the Stieltjes transform and the second to the Weierstrass transform. In both cases $\hat{K}$ decreases exponentially as $|\mathrm{Re}\, t| \to \infty$. There is an obvious reason for the restriction to kernels of such a decrease. If K decreases too slowly, it may happen that $\Psi(x) = 0$ for $x > 0$, does not imply $\phi(x) = 0$ for $x > x_0$. In such cases any remainder estimate has a minorant depending only on the kernel. This problem is evidently related to the uniqueness problem to be discussed in chapter 6 and we refer to example 4 in section 6.3.

The auxiliary function $\exp(-\pi(.)^2)$ employed in our method does not allow kernels of a very rapid decrease, but the main approach can be applied also with other auxiliary functions.

Our starting point is formula (2.2.1), where we put $\omega = \pi^2 \xi^{-1}$ and write

$$(1) \qquad \phi(x) = O(1)\Big\{ \sup_{0 \le z-y \le \Omega^{-1}} (\phi(x-y)E(y) - \phi(x-z)E(z)) + |T*\Psi(x)| \Big\} ,$$

where

$$(2) \qquad T(y) = \int \hat{K}(t)^{-1}\exp(2\pi i t y) \int_{|v|\leq \Omega} \exp(-\frac{\pi^2}{\xi}(t-v)^2 + 2\pi i v \eta)\hat{\chi}(v)(\frac{\pi}{\xi})^{\frac{1}{2}}\, dv\, dt .$$

Since ϕ is bounded, $E(y) = \exp(-\xi y^2)$ and $|E'(y)| \leq \xi^{1/2}$ we can rewrite the first term on the right in (1) and obtain

$$(3) \quad \phi(x) = O(1)\Big\{ \sup_{x-2\leq z\leq y\leq z+\Omega^{-1}\leq x+2} (\phi(y)-\phi(z)) + \exp(-\xi) + \Omega^{-1}\xi^{1/2} \sup_{|s|\leq 2} |\phi(x+s)| + |T*\Psi(x)| \Big\}.$$

If ξ is small it is more suitable to proceed as in section 2.3 and apply

$$(3') \quad \phi(x) = O(1)\Big\{ \sup_{2^{-1}x\leq z\leq y\leq z+\Omega^{-1}\leq 2x} (\phi(y)-\phi(z)) + \exp(-\frac{\xi}{4}x^2) + \Omega^{-1} + |T*\Psi(x)| \Big\}.$$

5.2 A remainder theorem for a class containing the Stieltjes transform

Theorem 5.2 below was first given in Ganelius 1964. The method has been extended by Frennemo 1966, 1967 to the d-dimensional Laplace transform which has a much more complicated behavior. Frennemo's results cover the theorem I published in Comptes Rendus 1956, which is the most comprehensive formulation of earlier results on the Laplace transform by Avakumović, Freud, Korevaar and me.

It is shown (cf. section 1.5 example 2) in Ganelius 1964 how to derive the following result for the Stieltjes transform from theorem 5.2 (cf. Vučković 1953, 1954).

Let ρ and ν be real numbers $\rho > \nu \geq 0$ and let r be an increasing function such that V defined by $V(x) = r(\exp x)$ fulfills $V(v) \leq qV(x)$ for $v \leq x+1$ and some q. Let σ be of locally bounded variation, $\sigma(0) = 0$ and suppose that

$$\int_0^\infty (\lambda+\omega)^{-\rho}\, d\sigma(\lambda) = O(\omega^{\nu-\rho})\exp(-r(\omega)) , \quad \omega \to \infty ,$$

and that

$$\sup_{\omega\leq\Omega\leq\omega+\omega/r(\omega)} \int_\omega^\Omega d\sigma(\lambda) \leq O(\omega^\nu/r(\omega)) , \quad \omega\to\infty .$$

Then

$$\sigma(\omega) = O(\omega^\nu/r(\omega)) , \quad \omega\to\infty .$$

We now introduce the sub-set E of $L(R)$ consisting of those kernels K for which there is a g holomorphic in $\mathrm{Im}\, t > -b$ and such that $g(t) = \hat{K}(t)^{-1}$ for real t and

(1) $$|g(t)| \leq M \exp 2\pi m|t|$$

for $\operatorname{Im} t > -b$.

THEOREM 5.2 :

Let $V : R \to R_+$ *be increasing and assume that there is a positive* q *such that* $V(v) \leq qV(x)$ *for* $v \leq x+1$. *Suppose that* $K \in E$ *and that* $\phi \in L^\infty$. *Then*

$$\psi(x) = K * \phi(x) = O(\exp(-V(x))) , \quad x \to \infty ,$$

and

(2) $$\sup_{x \leq y \leq x+1/V(x)} (\phi(y) - \phi(x)) = O(1/V(x)) , \quad x \to \infty ,$$

imply

$$\phi(x) = O(1/V(x)) .$$

For the proof we apply (5.1.2) and (5.1.3).

The first important step is to estimate T given by (5.1.2) under the assumptions of the theorem. We change the line of integration by the transformation $t \to t + i\zeta$, $\zeta \geq 0$ and introduce (1). We find that

$$|T(y)| \leq M \int \exp(2\pi m|t+i\zeta| - 2\pi\zeta y) \int_{|v| \leq \Omega} \exp(-\pi^2\xi^{-1}(t-v)^2 + \pi^2\zeta^2\xi^{-1}) \, dv \, (\pi/\xi)^{1/2} \, dt \leq$$

$$\leq M \exp(2\pi\zeta(m-y) + \pi^2\zeta^2\xi^{-1}) \int_{|v| \leq \Omega} dv \int \exp(2\pi m|t| - \pi^2\xi^{-1}(t-v)^2) \, dt \, (\pi/\xi)^{1/2} .$$

Substituting $t = v-u$, applying $|t| \leq |v| + |u|$ and observing that

$$\int \exp(2\pi mu - \pi^2\xi^{-1}u^2) \, du \, (\pi/\xi)^{1/2} = \exp(\xi m^2)$$

we obtain

$$T(y) = O(1) \exp(2\pi\zeta(m-y) + \pi^2\zeta^2\xi^{-1} + 2\pi m\Omega + \xi m^2) .$$

We choose $\pi\zeta = \xi(y-m)$ if $y \geq m$ and $\zeta = 0$ if $0 \leq y \leq m$.

If $y < 0$ we proceed similarly starting with a negative ζ. Taking $2\pi\Omega = m\xi$ we get

(3) $$T(y) = O(1) \exp(3\xi m^2 - \xi(m-|y|)^2) .$$

To compute $T * \psi$ we estimate $\psi(x-y)$ by $O(1)$ if $|y| \geq 3m$ and by $O(\exp(-V(x-3m)))$ if $|y| \leq 3m$. Thus

$$T*\psi(x) = \int\psi(x-y)T(y)\,dy = O(1)\Big\{\exp(3\xi m^2 - V(x-3m)) + \int_{|y|\geq 2m} \exp(3\xi m^2 - \xi y^2)\,dy\Big\} .$$

If $\xi > 1$ the last integral certainly is $O(\exp(-\xi m^2))$, and since $V(x-3m) > q^{-1-3m}V(x)$ we can choose $\xi = \delta V(x)$ and obtain

$$T*\psi(x) = O(\exp(-\varepsilon V(x)))$$

with $\varepsilon > 0$.

Recalling that $2\pi\Omega = m\xi$ and applying the tauberian condition (2), we find by insertion in (5.1.3) that

$$\phi(x) = O(1)\Big\{V(x)^{-1} + V(x)^{-1/2} \sup_{|s|\leq 2} |\phi(x+s)| + \exp(-\varepsilon V(x))\Big\} \tag{4}$$

and the boundedness of ϕ implies the preliminary result $\phi(x) = O(V(x)^{-1/2})$.

If this estimate is intoduced in (4) we get

$$\phi(x) = O(V(x)^{-1})$$

and the theorem is proved.

5.3 A best possible estimate for the Weierstrass transform

This time we consider the sub-set F of $L(R)$ consisting of those kernels K for which there is an entire g such that $g(t)\hat{K}(t) = 1$ for real t and $|g(t)| \leq M \exp(c|t|^2)$ for all complex t.

THEOREM 5.3.1 :

Let $V : R \to R_+$ *satisfy* $V(x) \leq V(kx) \leq k^2 V(x)$ *for all* $k \geq 1$. *Let* $K \in F$ *and* $\phi \in L^\infty$ *and assume that*

$$\psi(x) = K * \phi(x) = O(\exp(-V(x))) , \quad x \to \infty . \tag{1}$$

If

$$\sup_{x\leq y\leq x+V(x)^{-1/2}} (\phi(y)-\phi(x)) = O(V(x)^{-1/2}) , \quad x\to\infty , \tag{2}$$

then

$$\phi(x) = O(V(x)^{-1/2}) , \quad x\to\infty . \tag{3}$$

This theorem is best possible in the following sense.

THEOREM 5.3.2 :

If, for real t *and* ζ, *it holds that*

(4) $$|\hat{K}(t+i\zeta)| \leq M\exp(c\zeta^2-bt^2)$$

with $b > 0$ *and a* c *satisfying* $c\,\overline{\lim}\,x^{-2}V(x) < \pi^2$, *and if (1) and* $|\phi'| \leq 1$ *imply* $\phi(x) = O(1/Q(x))$, $x\to\infty$, *then*

$$Q(x) = O(V(x)^{1/2}), \quad x\to\infty .$$

(That $c \geq b$ is implied by the Phragmén-Lindelöf theorem. The importance of the number π^2 will be evident in next chapter.)

As previously we start by estimating T from (5.1.2). We introduce our estimate for $g(t) = \hat{K}(t)^{-1}$ and change the line of integration by $t\to t+i\zeta$. Thus

$$T(y)=\int g(t+i\zeta)\exp(2\pi ity-2\pi\zeta y)\int\hat{\chi}(v)\exp(2\pi i\eta v-\pi^2\xi^{-1}(v-t-i\zeta)^2)(\pi/\xi)^{1/2}\,dv\,dt =$$

$$= O(\exp(-2\pi\sigma y+\pi^2\zeta^2\xi^{-1}+c\zeta^2)\int_{|v|\leq\Omega}dv\int\exp(ct^2-\pi^2\xi^{-1}(v-t)^2)(\pi/\xi)^{1/2}\,dt .$$

We fix $\xi > c^{-1}\pi^2$. The inner integral can be computed and gives $C\exp(av^2)$ for some C and a. Hence

$$T(y) = O(\exp(-2\pi\zeta y+2c\zeta^2+a\Omega^2)) = O(\exp(a\Omega^2-dy^2)) ,$$

for some d by taking $\zeta = (2c)^{-1}\pi y$. We next observe that

(5) $$T*\Psi(x) = O(\exp(a\Omega^2))\int\exp(-V(y)-d(x-y)^2)dy = O(\exp(a\Omega^2-\gamma V(x))) ,$$

since our assumptions on V imply that $V(y) \geq 4^{-1}V(x)$ for $y \geq \frac{1}{2}x$ and that there is a γ such that

$$\int_{|y-x|>\frac{1}{2}x}\exp(-d(x-y)^2)dy = O(\exp(-\tfrac{1}{4}dx^2)) = O(\exp(-\gamma V(x))) .$$

Since we have fixed ξ we apply (5.1.3') and get

$$\phi(x) = O(1)\Big\{\sup_{\frac{1}{2}x\leq z\leq y\leq z+\Omega^{-1}\leq 2x}(\phi(y)-\phi(z)) + \exp(-cx^2) + |T*\Psi(x)| .$$

If we recall (2) and (5) we see that $2a\Omega^2 = \gamma V(x)$ gives

$$\phi(x) = O(1)\left\{V(x)^{-1/2}+\exp(-\tfrac{1}{2}\gamma V(x))\right\} = O(V(x)^{-1/2}) \ .$$

We next turn to the proof of theorem 5.3.2 which follows the same lines as the proof of 3.2.2.

Our theorem tells us that the norms

$$|||\phi|||_1 = \|\phi\|_\infty + \|\phi'\|_\infty + \|(K*\phi)\exp V\|_\infty$$

and

$$|||\phi|||_2 = |||\phi|||_1 + \|Q\phi\|_\infty$$

are eqaivalent. Hence, as previously,

$$\delta \sup|Q(x)\phi(x)| \le \|\phi\|_\infty + \|\phi'\|_\infty + \|(K*\phi)\exp V\|_\infty \ .$$

We take

$$\phi(x) = \exp(2\pi i\Omega(x-y)-\xi(x-y)^2)$$

so that

$$\hat{\phi}(t) = \exp(-2\pi ity-\pi^2\xi^{-1}(t-\Omega)^2)(\pi/\xi)^{1/2} \ .$$

Hence

$$Q(y) = O(1)\left\{1+\Omega+\xi^{1/2}+\|(K*\phi)\exp V\|_\infty\right\} . \tag{6}$$

We consider

$$K*\phi(x) = \exp(2\pi it(x-y) - \pi^2\xi^{-1}(t-\Omega)^2)\hat{K}(t)\,dt(\pi/\xi)^{1/2} \ .$$

If we apply (4) after the change $t \to t+i\zeta$, we get

$$K*\phi(x) = O(1)\exp(-2\pi\zeta(x-y)+\pi^2\zeta^2\xi^{-1}+c\zeta^2)\int\exp(-bt^2-\pi^2\xi^{-1}(t-\Omega)^2)(\pi/\xi)^{1/2}\,dt \ .$$

If we fix ξ the contribution from the integral is $\exp(-\gamma\Omega^2)$ and if ξ is large we get

$$K*\phi(x) = O(1)\exp(-2\pi\zeta(x-y) + (c+\varepsilon)\zeta^2 - \gamma\Omega^2) \ .$$

Putting $(c+\varepsilon)\zeta = \pi(x-y)$ we find that

$$K*\phi(x) = O(1)\exp(-\gamma\Omega^2 - \eta(x-y)^2) \ ,$$

for any $\eta < \pi^2c^{-1}$ with $\gamma = \gamma(\eta)$.

Our next problem is to estimate $\|(K * \phi)\exp V\|$. We consider

$$W(x) = V(x) - \eta(x-y)^2 .$$

Recalling the assumptions on V, we see that

$$W(ky) = V(ky) - \eta(k-1)^2y^2 \leq k^2V(y) - \eta(k-1)^2y^2 .$$

The number η can be chosen so that $\overline{\lim}\ y^{-2}V(y) < \eta$. Let us only consider y-values such that $V(y) < (\eta-\varepsilon)y^2$ for some ε. Fix y. It follows that $W(ky) < 0$ for k larger than some finite number, and, since $W(x) \leq V(x) \leq V(ky)$ for $x \leq ky$, it holds that $W(x) \leq CV(y)$ for all x. Hence

$$\|(K * \phi)\exp V\| = O(1)\exp(-\gamma\Omega^2 + C\,V(y)) \quad ,$$

and insertion in (6) gives

$$Q(y) = O(V(y)^{1/2}) ,$$

by taking

$$\gamma\Omega^2 = 2CV(y) .$$

CHAPTER 6

CONVOLUTION EQUATIONS WITHOUT NON-TRIVIAL SOLUTIONS

6.1 Introduction

In the introduction to the preceding chapter I mentioned that for certain kernels there is a critical rate of decrease. If the remainder decreases more rapidly we can conclude that it is identically zero. The best known example is the Laplace transform and the special case of power series, which has been studied by several authors.

The following precise formulation is given by Korevaar 1954 b.

If $a_n = O(n^p)$ for some real p and if

$$f(x) = \sum_0^\infty a_n x^n = O(\exp(-\omega(x)(1-x)^{-1}),\ x \uparrow 1,$$

with

$$\overline{\lim_{x \uparrow 1}}\ \omega(x) = \infty,$$

then $a_n = 0$ for all n.

If Korevaar's result is transformed to convolution form we get that the critical rate is $\psi(x) = \exp(-\exp x)$, if $\hat{K}(t) = \Gamma(1+2\pi it)$. In chapter 5 I mentioned the critical rate $\psi(x) = \exp(-\pi x^2)$ for the Weierstrass transform with $\hat{K}(t) = \exp(-\pi t^2)$. Some results of this kind are discussed in Hirschman-Widder 1955, p.247 f.

I am going to show that these results and several others follow from some known results for Fourier transform by the distributional method introduced in section 1.4 .

A pair (η,ζ) of positive-valued functions on the real axis will be called a trivial pair for Fourier transforms, if for $f \in L(R)$,

$$f(x) = O(\exp(-\eta(x))),\ x \to \infty,$$

and

$$\hat{f}(t) = O(\exp(-\zeta(t))),\ |t| \to \infty,$$

imply that $f = 0$.

Characterizations of trivial pairs have been given by Hardy 1933, Morgan 1934, Levinson 1940 and Hirschman 1950 among others. Levinson's main theorem is as follows:

If $\eta'' > C > 0$, and the inverse of η' is H, then (η,ζ) is a trivial pair, if

(1) $$\overline{\lim_{x\to\infty}} \int_{x \geq |y| \geq 1} (y^{-2} - x^{-2})\, \zeta(y)dy - 2\pi^2 H(x) = \infty .$$

This result is precise in the neighborhood of the classical result of Paley-Wiener which says that if $f \in L^2(R)$, $f(x) = 0$ for $x > x_0$ and

$$\int_{-x}^{x} y^{-2}\zeta(y)\, dy = \infty ,$$

then $f = 0$.

It is not precise e.g. in the case studied by Hardy where both $\zeta(x)$ and $\eta(x)$ are of the order x^2. Levinson also gives a method to handle these cases and it is not difficult to formulate a general relation. This has been done by Gustavsson 1966, but I don't want to go into details about trivial pairs. A second theorem of Levinson's reads as follows.

If $\eta(x) = Ax^p$, $p \geq 2$, *and* $\zeta(t) = Bt^q + r(t^q)$, $t \to +\infty$, *with* $p^{-1}+q^{-1} = 2$, *and*

(2) $$B(q)^p(Ap)^q = (2\pi)^{pq}(\sin \tfrac{\pi}{p})^p ,$$

then r *increasing with*

(3) $$\int_1^\infty u^{-2}r(u)\,du = \infty ,$$

implies that (η,ζ) *is a trivial pair.*

6.2 Trivial convolution equations

We now turn to our theorem (of section 1.5).

THEOREM 6.2 *Let* $K \in O_C'$ and $\phi \in \mathcal{S}'$, *and* $\psi = K * \phi$.

If (η,ζ) *is a trivial pair and* $|D^m\hat{K}(t)| = O(\exp(-\zeta(t))$ *for all positive integers* m, *and* ψ *is a function fulfilling*

$$\psi(x) = O(\exp(-\eta(x))),$$

then $\psi = 0$. *Hence, if* $\hat{K}(t) \neq 0$, *it follows that*

$$\phi = 0 .$$

The proof is as follows.

Take $h \in \mathcal{S}$ such that $\hat{h} \in \mathcal{D}$, and consider

$$(\psi h)^\wedge(t) = \hat{\psi}(\hat{h}(t-.)) = \hat{K}\hat{\phi}(\hat{h}(t-.)) = \hat{\phi}(\hat{K}\hat{h}(t-.)).$$

The structure of tempered distributions being well-known we introduce

$$\hat{\phi} = D^m F$$

with a tempered function F . We find that

$$(\psi h)^{\wedge}(t) = \int F(t)\,(\sum D_{\tau}^{\nu}\hat{K}(\tau)D^{m-\nu}\hat{h}(t-\tau))\,d\tau,$$

and our assumptions imply that

$$(\psi h)^{\wedge}(t) = O(1)\max_{|u|\leq C}\exp(-\zeta(t+u)).$$

Since translation in a Fourier transform is equivalent to multiplication of the original function with a factor of modulus 1, we have a pair of a function ψh and its Fourier transform and they are dominated by a trivial pair.

Hence ψh is zero for all h with $\hat{h}\in\mathcal{D}$, and it follows that $\psi = 0$. The proof is evidently valid in any number of dimensions.

6.3 Applications

Let us now see that the theorem given implies the results mentioned previously.

Example 1: If $\hat{K}(t) = \exp(-\pi t^2)$, our theorem and formulas (6.1.2) and (6.1.3) show that $\psi(x) = \exp(-\pi x^2)$ has the critical rate of decrease. If $p = q = 2$ formula (6.1.2) reads $AB = \pi^2$.

Example 2: The Laplace transform has $K(x) = \exp(-(p+1)x-\exp(x))$, so that $K(t) = \Gamma(1+p+2\pi it)$. Hence we can take $\zeta(t) = \pi^2|t| - (p+1)\log(1+|t|)$. Applying Levinson's theorem we see by (6.1.1) that

$$\overline{\lim}\,(\log x - H(x)) = \infty$$

is the condition for triviality. But if $\eta(x) = \omega(x)e^x$ with a differentiable ω satisfying $\overline{\lim}\,\omega(x) = \infty$ then

$$\eta'(x) = \exp(x+\chi(x))$$

where $\overline{\lim}\,\chi = \infty$. Hence the inverse H satisfies

$$H(x) = \log x - \sigma(x),\ \overline{\lim}\,\sigma = \infty .$$

To apply the theorem we need some regularity assumptions on ω , but we get the substantial part of Korevaar's theorem .

Example 3: The Stieltjes transform with $K(x) = (1+\exp x)^{-2}\exp(\delta x)$ has $\hat{K}(t) = O(|t|\exp(-2\pi^2|t|))$. We apply Levinson's theorem again and find that $\overline{\lim}\,(2\log x - H(x)) = \infty$. It follows that the critical rate is given by $\psi(x) = \exp(-c\exp(\frac{1}{2}x))$. (A slightly weaker result is given in Vučković 1953. Cf. also Hirschman-Widder , p.247).

The Paley-Wiener theorem may be applied to characterize kernels for which ψ may be 0 on an interval (x_0,∞) without the same holding for ϕ . Let us just mention the following example.

Example 4: There is a $K \in L(R)$ such that $\hat{K}(t) = (1 + |t|^p \operatorname{sgn} t)^{-1}$, $\frac{1}{2} < p < 1$. We can find a bounded ϕ with bounded ϕ' for large values such that $K * \phi(x) = 0$ for $x > 0$ but $\overline{\lim_{x \to \infty}} \; x^{1-p} |\phi(x)| > 0$.

This follows by computation if $K * \phi$ is taken to be the characteristic function of a bounded interval on the negative axis.

CHAPTER 7

COMPLEX TAUBERIAN THEOREMS

7.1 Introduction

In the problems we have considered so far, the assumptions concern the behavior of the transform on the real axis. There are many interesting tauberian theorems, in which information about 'the convolution transform' is provided in the complex plane, the best known probably being Fatou's theorem and Ikehara's theorem.

If we consider a power series $f(z) = \sum_0^\infty a_n z^n$ convergent in the unit circle, Littlewood's theorem tells us that the existence of the limit $\lim_{x\uparrow 1} f(x) = 0$ implies convergence of the series to 0 if $na_n = O(1)$. Fatou's theorem shows that regularity of the function at $z=1$ implies convergence if $a_n = o(1)$. If we assume regularity of the function with $f(1)=0$, and $na_n = O(1)$, then we have the remainder estimate $s_n = \sum_{m\leq n} a_m = O(1/n)$, which is a special case of a theorem proved by Korevaar 1954 b.

If assumptions about regularity at boundary points are invoked, it is easy to understand that the Wiener approach is not very suitable. However there are theorems of an intermediate character, where our methods can be applied to give new interesting results. The following theorem is typical and is found among other similar results in Subhankulov 1964.

Let the integral

$$f(s) = \int_0^\infty \exp(-us)b(u)\, du , \quad s = \sigma + it ,$$

converge absolutely for $\mathrm{Re}\, s > 0$ and assume that

$$(1) \qquad f(s) = O(|s|^{-\alpha}) , \quad \alpha \geq 0 ,$$

in a domain containing the positive axis and with a boundary near the point $s=0$ given by $|t| = \sigma^\gamma$, $0<\gamma\leq 1$. If $b(u) = O(u^\lambda)$, $\lambda \geq 0$, then

$$B(x) = \int_0^x b(u)du = O(x^{\lambda+\gamma}) + O(x^\alpha) .$$

If we assume (1) on the real axis only, then we only get $O(x^{\lambda+1}(\log x)^{-1})$ instead of $O(x^{\lambda+\gamma})$.

A related theorem for the Stieltjes transform has been proved by Malliavin 1962 and Pleijel 1963. That theorem is stated in section 7.2 and shown to be a special case of a general complex remainder theorem.

For the special theorem Pleijel's method is certainly simpler, but it depends strongly on the special properties of the Stieltjes transform. The method I use has been extended by A.Lydén to a class of kernels containing the Laplace transform, and I state his theorem, which generalizes the above mentioned result of Subhankulov.

It should be pointed out that in most applications of tauberian theorems to number theory, information in the complex plane is provided. Subhankulov's work aims at such applications.

Several of the real tauberians for Laplace and Stieltjes transform have been developed for applications in the spectral theory of elliptic differential operators (cf. e.g. Bergendal 1956). The last years better estimates for the relevant functions have been obtained within the theory of differential operators and made the applications of complex tauberian theorems possible. Thus the theorem of Malliavin-Pleijel just mentioned was required for such applications by Agmon-Kannai 1965.

The final form of the theory of asymptotic properties of the spectral functions just mentioned given by Hörmander 1966 avoids reference to tauberian theorems. Their place is taken by some precise lemmas related to Hörmander's generalization of Bohr's inequality from 1954.

Results of that type are very useful in the proofs of those complex remainder tauberians where the general method fails. My original approach to the general remainder theorems started with an application of a special case of Bohr's inequality (cf. Ganelius 1962)and it is disguised in the method of chapter 2.

In section 7.3 I prove an elementary local version of Bohr's inequality which I have used in lectures to obtain remainder variants of Fatou's and Ikehara's theorem. Some simple applications of that kind conclude the chapter.

7.2 A theorem with information in the complex plane

We restrict the class E considered in section 5.2 by adding the condition that to every real ζ there is a M_ζ so that

$$\int_{-\infty}^{\infty} |g(t+i\zeta|^2 \exp(-4\pi m|t|)dt < M_\zeta .$$

Let us call the new class E_2.

THEOREM 7.2 :

Let $U:R\to R_+$ *and* $V:R\to R_+$ *be decreasing to* 0 *and assume that there is a positive* q *such that*

$$U(v) \geq qU(x), \quad V(v) \geq qV(x) \quad \text{for} \quad v\leq x+1 .$$

Suppose that $K \in E_2$, $\phi \in L^\infty$ _and that_

(1) $$\sup_{x \leq y \leq x+U(x)} (\phi(y) - \phi(x)) = O(U(x)) , \quad x \to \infty .$$

_If_ $\psi = K * \phi$ _and_ ψ _has a holomorphic continuation in the domain._

(2) $$S = \{ \xi + i\eta : |\eta| \leq m - U(\xi) \} ,$$

such that

(3) $$\psi(\xi + i\eta) = O(V(\xi)) , \quad \xi \to \infty , \quad \xi + i\eta \in S,$$

then

$$\phi(x) = O(U(x)) + O(V(x)) , \quad x \to \infty .$$

This theorem is a generalization of results of Malliavin 1962 and Pleijel 1963. The main part of their theorem reads as follows.

Let $d\sigma$ _be a positive measure on the non-negative axis, and put_ $f(x) = \int_0^\infty (\lambda + z)^{-1} d\sigma(\lambda)$. _Let_ L _be the curve_ $z = -x \pm ix^\gamma$, $0 < x < \infty$, _where_ $0 \leq \gamma < 1$, _and assume that_

$$f(z) = az^{-\alpha} + O(z^{-\beta}) , \quad L \ni z \to \infty ,$$

where $1 > \beta > \alpha > 0$. _Then it holds that_

$$\int_0^x d\sigma(\lambda) = Ax^{1-\alpha} + O(x^{\gamma - \sigma}) + O(x^{1-\beta}), \quad x \to \infty .$$

A Phragmén-Lindelöf argument and an integration by parts in the integral for f, followed by the standard transformations for bringing the Stieltjes transform in convolution form gives a formulation covered by our theorem with $U(x) = \exp(-(1-\gamma)x)$, $V(x) = \exp(-(\beta - \alpha)x)$ and

$$g(t) = \hat{K}(t)^{-1} = \frac{\sinh(2\pi^2 t + i\pi\alpha)}{i\pi(1-\alpha) - 2\pi^2 t}$$

Evidently $K \in E_2$ with $m = \pi$.

We now turn to the proof of theorem 7.2. We shall apply the same method as in section 5.2 and we recall the estimate (5.2.3) which we shall use if $|y| \geq 4m$. In the remaining interval we shall improve the estimate for T by moving out into the complex plane.

We write $T(y) = T_+(y) + T_-(y)$, where

$$T_+(y) = \int_0^\infty g(t)\exp(2\pi ity) \int_{|v|\leq\Omega} \exp(-\pi^2\xi^{-1}(v-t)^2+2\pi iv\eta)dv(\pi/\xi)^{1/2}dt ,$$

and correspondingly for T_- . Evidently T_+ is a holomorphic function of y and we shall estimate $T_+(y+i\Theta)$ by transforming the line of integration into the contour $t = is$, $0\leq s\leq\zeta$, $t = i\zeta+u$, $0<u<\infty$. We get

$$T_+(y+i\Theta) = O(1)\int_0^\zeta |g(is)|\exp(-2\pi ys) \int_{|v|\leq\Omega} \exp(-\pi^2\xi^{-1}(v^2-s^2))ds(\pi/\xi)^{1/2} dv +$$

$$+ O(1)\exp(-2\pi y\zeta) \left|\int_0^\infty \exp(2\pi iyu-2\pi u\Theta)g(u+i\zeta) \int_{|v|\leq\Omega} \exp(-\pi^2\xi^{-1}(u+i\zeta-v)^2+2\pi iv\eta)dv(\frac{\pi}{\xi})^{\frac{1}{2}}du\right| =$$

$$= I_1 + I_2 .$$

Now

$$I_1 = O(1)\int_0^\zeta \exp(2\pi s(m-y) + \pi^2\xi^{-1}s^2) ds ,$$

and we observe that $I_1 = O(1)$, if $y\geq 3m$ and $\pi\zeta\leq\xi(y-m)$, and also that $I_1 = O(1)$ for all bounded y if ζ is fixed.

If $\Theta\geq 0$, we can estimate I_2 in exactly the same way as in section 5.2 and obtain an estimate like the one given in (5.2.3), that is

$$I_2(y+i\Theta) = O(1)\exp(3\xi m^2 - \xi(m-|y|)^2) =$$

$$= O(1)\exp(-\xi m^2) ,$$

for $|y|\geq 3m$.

If $|y|\leq 3m$ we only consider $\Theta = m - \Omega^{-1}$ and get

$$I_2(y + im - i\Omega^{-1}) =$$

$$= O(1)\int_0^\infty \exp(2\pi iyu+2\pi u\Omega^{-1})g(u+i\zeta)\exp(-2\pi mu)\int_{|v|\leq\Omega} \exp(-\pi^2\xi^{-1}(u+i\zeta-v)^2+2\pi iv\eta)dv(\frac{\pi}{\xi})^{\frac{1}{2}}du.$$

We observe that if ζ is finite and $\pi\Omega>\xi$, then the inner integral is $O(1)\exp(-\pi(|u|-\Omega))$ for $|u|\geq 2\Omega$, and $O(1)$ for all u. Hence

$$\int_{-3m}^{3m} |I_2(y+im-i\Omega^{-1})|^2 dy =$$

$$= O(1)\int_0^\infty |g(u+i\zeta)\exp(-2\pi mu)|^2 du(1+\max_{|u|\geq 2\Omega}\exp(2\pi u\Omega^{-1}-\pi(|u|-\Omega))) = O(1) .$$

Summing up, we have proved that

$$T_+(y+i\zeta) = O(1)$$

for all non-negative Θ and all $|y|\geq 3m$, and that

$$\int_{-3m}^{3m} | T_+(y+im-i\Omega^{-1})|^2 dy = O(1) .$$

Similar estimates hold for $T_-(y+i\zeta)$.

To estimate $T*\psi(x)$ we now change the line of integration in the following way.

$$T*\psi(x) = \int_{|y|\geq 4m} \psi(x-y)T(y)dy + \int_{|y|\leq 3m} \psi(x-y-im+i\Omega^{-1})T_+(y+im-i\Omega^{-1})dy +$$

$$+ \int_{|y|\leq 3m} \psi(x-y+im-i\Omega^{-1})T_-(y-im+i\Omega^{-1}dy + \int_{3m\leq|y|\leq 4m} \psi(x-y-i\Theta(y))T_+(y+i\Theta(y))dy+$$

$$+ \int_{3m\leq|y|\leq 4m} \psi(x-y+i\Theta(y))T_-(y+i\Theta(y))dy ,$$

where Θ is a linear function of y making the path of integration continuous.

If we apply our estimates to the integrals in the order given above we get

$$T*\psi(x) = O(\exp(-m^2\xi)) +(\int_{|y|\leq 3m} |\psi(x-y-im+i\Omega^{-1})|^2 dy)^{1/2} +$$

$$+ (\int_{|y|\leq 3m} |\psi(x-y+im-i\Omega^{-1})|^2 dy)^{1/2} + O(V(x-4m)) + O(V(x-4m)) .$$

If $\Omega^{-1} = \delta U(x)$, the L_2-norms are of the order $V(x)$ and if $\xi = \mathcal{E} V(x)$, then

$$T*\psi(x) = O(U(x)) + O(V(x)) .$$

Applying (5.1.3) we get

$$\phi(x) = O(1) \left\{U(x) + U(x)^{1/2} \sup_{|s|\leq 2} |\phi(x+s)| + V(x)\right\} .$$

Iterating as in the proof of theorem 5.2 we get

$$\phi(x) = O(U(x)) + O(V(x))$$

and theorem 7.2 is proved.

The theorem of Lydén 1970 deals with kernels in the class $E(m)$. An integrable function K belongs to $E(m)$, if there is an entire function g such that $g(t)\hat{K}(t) = 1$ for real t and g fulfills the following three conditions.

There are positive numbers m, n, r, $k \geq 1$ and C, such that

(i) $\quad |g(t+iu)| \leq C\exp(m|t| + u\log(1+|u|) + r|u|)$, $\quad u \leq k$,

(ii) $\quad |g(t+iu)| \leq C\exp(n|t| + u\log u + ru)$, $\quad k < u$,

(iii) $\quad \int_{-\infty}^{\infty} |g(t+ik)|^2 \exp(-2m|t|)\, dt \leq C$.

For functions in this class we have the following theorem.

<u>Suppose that</u> $K \in E(m)$ <u>and that</u> ϕ <u>and</u> ϕ' <u>belong to</u> L^∞. <u>If</u> $\psi = K * \phi$ <u>has a holomorphic continuation in the domain</u>

$$S = \{x + iy : |y| \leq m - \exp(-(\beta-1)x)\}, \quad \beta > 1,$$

<u>such that</u>

$$\psi(x+iy) = O(\exp(-ax)), \quad x + iy \in S, \quad x \to \infty, \quad 0 \leq a \leq k ,$$

<u>then</u>

$$\phi(x) = O(\exp((\beta^{-1}-1)x)) + O(\exp(-ax)) .$$

The application to Subhankulov's theorem is straightforward. The kernel related to the Laplace transform belongs to $E(\pi/2)$, and we put $\gamma = \beta^{-1}$.

7.3 <u>A localized version of Bohr's inequality</u>

For completeness I want to include remainder theorems corresponding to the classical complex tauberian theorems, i.e. Fatou's and Ikehara's. They will both be proved by a variant of Bohr's inequality. As mentioned in the introduction deeper and more general theorem are known (Hörmander 1954, 1966), but the following is sufficient for our present purposes.

We shall deal with a class $\mathfrak{M}$ containing those functions $m: R \to [1,\infty)$ which are non-decreasing and satisfy

$$m(2x) \leq Cm(x) \quad \text{if} \quad x \geq 0 ,$$

$$m(x) = 1 \quad \text{if} \quad x \leq 0 .$$

It is immediately seen that every function in $\mathcal{M}$ is dominated by a polynomial.

THEOREM 7.3 :

Let $g \in L^{\infty}$ and assume that

(1) $$\sup_{x \leq y \leq x+1/T} (g(y) - g(x)) \leq K/m(x) ,$$

with a $m \in \mathcal{M}$.

If $\hat{g}(t) = 0$ for $|t| < T$, then

(2) $$|g(x)| \leq MK/m(x) ,$$

where M only depends on m and is independent of T.

Without loss of generality we take $T = 1$, since the general case follows from the special by considering $g(./T)$.

We fix a function $h \in \hat{\mathcal{D}}$ with $0 \leq h(x) \leq h(0)=1$, supp $\hat{h} \subset (-1/2, 1/2)$ and $|h'(x)| \leq 1$. Then there is a constant M_1 such that $|h(x)| \leq M_1/m(x)$. With $0<r<1$ we form $h_r(x) = h(rx)$, so that $\hat{h}(t) = r^{-1}h(r^{-1}t)$. Considering $u = gh_r(x-.)$ we find that

$$\hat{u}(t) = (gh_r(x-.))^{\wedge}(t) = \hat{g}(\hat{h}_r(.-t)\exp(2\pi ix(.-t))) = 0$$

for $|t| < 1/2$ according to our assumptions on the support of $\hat{g}$ and $\hat{h}$.

Thus, if χ is the function defined in section 2.1 with Ω taken $= 1/2$, then $u * \chi = 0$. Proceeding exactly as in the proof of theorem 2.1 we get

(3) $$|g(x)| \leq \sup_y |g(y)h_r(x-y)| \leq 4 \sup_{0 \leq v-u \leq 4} (g(v)h_r(x-v) - g(u)h_r(x-u)) .$$

As usual we rewrite

$$g(v)h_r(x-v) - g(u)h_r(x-u) = (g(v)-g(u))h_r(x-v) + g(u)(h_r(x-v)-h_r(x-u))$$

and since $|h_r'(x)| \leq r$ we obtain

(4) $$|g(x)| \leq 4 \sup_{u \leq v \leq u+4} \left\{4K/m(u) + 4r|g(u)|\right\} \leq 16 \sup_u (K/m(u) + r|g(u)|) .$$

We shall use this formula for $u > x/2$. If $u \leq x/2$, $g(u)h_r(x-u)$ and $g(v)h_r(x-v)$ are dominated by

$$\|g\|_\infty M_1/m(rx/2) \leq M_r \|g\|_\infty/m(x) ,$$

where M_r depends on m and r. Now $\|g\|_\infty \leq 16K$ as can be seen since $m \geq 1$, if we let $r \to 0$ in(3). Adding the estimates given by (3) for $u \leq x/2$ and by (4) for $u > x/2$ and recalling that $\sup_{u>x/2}(1/m(u)) \leq C/m(x)$ we get

$$(5) \qquad |g(x)| \leq 8 \cdot 16KM_r/m(x) + 16KC/m(x) + 16r \sup_{u>x/2} |g(u)| .$$

We now take r so that $16rC = 1/2$ and rewrite (5) as

$$(6) \qquad |g(x)| \leq (1/2)KM/m(x) + (1/2C) \sup_{u>x/2} |g(u)| .$$

In particular

$$|g(u)| \leq (1/2)KM/m(u) + (1/2C) \|g\|_\infty ,$$

and if we apply that estimate in (6) we get

$$|g(x)| \leq (1/2)KM/m(x) + (1/4C)KMC/m(x) + (1/4C^2) \|g\|_\infty .$$

Since $C \geq 1$ we get by iteration that

$$|g(x)| \leq KM(2^{-1}+\dots,+2^{-N})/m(x) + 2^{-N} \|g\|_\infty ,$$

and hence

$$|g(x)| \leq KM/m(x).$$

7.4 Korevaar's extension of Fatou's theorem

The following theorem is essentially due to Korevaar 1954 b.

THEOREM 7.4 :

Suppose that $f(z) = \sum_0^\infty a_k z^k$ is regular at the point $z = 1$ on the circle of convergence and that $m(n)a_n \geq -1$ with $m \in \mathcal{M}$. Then $\sum a_k$ converges and $m(n)\sum_{n+1}^\infty a_k = O(1)$.

Putting $z = \exp(-2\pi\zeta)$ we obtain

$$F(\zeta) = f(\exp(-2\pi\zeta)) = \sum_0^\infty a_k e^{-2\pi k\zeta} ,$$

so that

$$F(s+it) = \int_0^{\infty} e^{-2\pi itx} d\alpha(x) = 2\pi it\hat{\alpha}(t) + F(s) ,$$

where

$$\alpha(x) = -\sum_{x \leq k} a_k e^{-2\pi ks} = \sum_{k<x} a_k e^{-2\pi ks} - F(s) .$$

Since $\hat{\alpha}(t) = (2\pi it)^{-1}(F(s+it)-F(s))$ and F is a regular analytic function at $s = 1$, we can find a function r such that $\hat{r}(t) = \hat{\alpha}(t)$ for small t and $|r(x)| \leq C_n(1+|x|)^{-n}$ for any n and independently of s. The function $g(x) = \alpha(x) - r(x)$ satisfies the conditions of theorem 7.3 for some T and if n is large enough we get $\alpha(x) = O(1/m(x))$. Hence

$$\left|\sum_{k<x} a_k e^{-2\pi ks} - F(s)\right| = |\alpha(x)| \leq C/m(x) ,$$

and

$$\sum_{k \leq x} a_k = f(1) + O(1/m(x)) .$$

7.5 Ikehara's theorem

In section 7.4 the Fourier transform of the function considered was supposed to be analytic and regular at the origin. We now assume that the Fourier transform has a modulus of continuity ω for $|t| \leq T$.

THEOREM 7.5 :

Let $g \in L^{\infty}$ and assume that

$$\sup_{x \leq y \leq x+1/T} (g(y) - g(x)) \leq K/m(x) .$$

If the L_1-modulus of continuity of g on $[-T,T]$ is dominated by a function ω such that $\omega(1/.) \in \mathcal{M}$, then

$$|g(x)| = O(1/m(x)) + O(\omega(1/x)) , \quad x \to \infty .$$

We continue $\hat{g}$ to function $\hat{h}$ with compact support and the same modulus of continuity. Then

$$h(x) = \int \exp(2\pi itx)\hat{h}(t)\, dt = 1/2 \int \exp(2\pi itx)(\hat{h}(t)-\hat{h}(t+1/2x))\, dx = O(\omega(1/x)) ,$$

and applying theorem 7.3 to $g(x) - h(x)$ we obtain our theorem.

The theorem can be reformulated as a tauberian remainder theorem for $f(z) = \int_0^{\infty} e^{-zx} g(x)\, dx$ with assumptions on the modulus of continuity of f on

$\{it: |t| \leq T\}$.

Ikehara's theorem reads as follows.

Let a be nondecreasing and assume that

$$A(s+it) = \int_0^\infty e^{-y(s+it)} da(y)$$

converges absolutely for $s \geq 0$. If f defined by $f(z) = A(z) - (z-1)^{-1}$ is continuous in $s \geq 1$, then $\lim_{y\to\infty} a(y)e^{-y} = 1$.

The proof we give is essentially the well-known simple proof of Bochner.

An integration by parts and introduction of $b(y) = e^{-y}a(y) - 1$ gives

$$A(z) = z\int_0^\infty e^{-y(z-1)} b(y)\, dy + z/(z-1) .$$

Now

$$b(y+h) - b(y) = e^{-(y+h)}(a(y+h) - a(y)) + a(y)e^{-y}(e^{-h} - 1) \geq -Kh ,$$

if we assume that $e^{-y}a(y) = O(1)$, which can be seen in various ways. E.g. theorem 7.3 can be used to obtain a bound for $\sup|e^{-sy}a(y)|$ independent of $s > 1$.

We now apply theorem 7.5 with $m(x) = T$ and get

$$|b(x)| \leq C(T^{-1} + \omega_T(1/x)) ,$$

where ω_T is the L_1-modulus of continuity of $\hat{b}$ on $(-T, T)$. We have $\hat{b}(t/(2\pi)) = (1+it)^{-1}A(1+it)-(it)^{-1}$. Hence $\overline{\lim}|b(x)| \leq CT^{-1}$ for every T and hence $\lim_{x\to\infty} e^{-x}a(x) = 1$.

CHAPTER 8

MISCELLANEOUS REMARKS

8.1 Approximation theorems. Wiener's second tauberian theorem

As we have seen in the first chapter we need an estimate for the degree of approximation, if we want to prove remainder theorems by approximation methods. Some results of this kind are given in Ganelius 1969. As an example we quote the following theorem, by aid of which precise remainder estimates can be obtained in many problems concerning the Laplace transform. Since the proof uses the devices applied in chapters 2-5, it is omitted.

THEOREM 8.1 :

Let $h \in L(R)$, $g \in L(R)$. *Assume that* $k = 1/\hat{h}$ *can be analytically continued to an entire function satisfying the inequality.*

$$|k(t+i\sigma)| \leq C\exp(\sigma \log(1+|\sigma|) + s|\sigma| + r|t|) .$$

Assume further that $D\hat{g}(t) = O(1)(1 + |t|)^{-m-1}$ *on the real axis.*

Then there is a positive q *such that every* $\Omega > 1$ *we can find a* $u \in L$ *with*

$$\|u\| < \exp\Omega, \qquad \|u * h * g - g\| = O(\Omega^{-m}) ,$$

and

$$\operatorname{supp} u \subset [\log\Omega - q, \log\Omega + q] .$$

Wiener's formulation of his general tauberian theorem includes a second variant dealing with the sub-class M of L of functions for which

$$\sum_{n=-\infty}^{\infty} \sup_{0 \leq x \leq a} |f(b + na + x)| < \infty .$$

Remainder theorems in this form can either be proved directly or deduced from our previous theorems. I quote the following result by Bergström 1971.

Let $H \in M$ *and assume that* $H^{(m-2)}$ *is the integral of a function of bounded variation. (If* $m = 1$, *then* H *is supposed to be of bounded variation.) Let* $K \in M$. *Assume that* $1/\hat{K}$ *and* $\hat{H}/\hat{K}$ *can be continued as holomorphic functions in a strip* $|\operatorname{Im} t| < c$ *and that they are* $O(\exp(q|t|^{\alpha}))$ *in that strip for some* $q > 0$ *and* α *with* $0 < \alpha \leq 2$. *Let* g *be of bounded varia-*

tion and suppose that $\int_x^{x+t} |dg(y)| \leq Ct$ for all x and positive t. Then

$$K * dg(x) = O(\exp(-W(x))) , \quad x \to \infty ,$$

implies that

$$H * dg(x) = O((W(x))^{-m/\alpha}) , \quad x \to \infty ,$$

for any sub-additive W with $\int \exp(W(x) - (c-\varepsilon)|x|)dx < \infty, \varepsilon > 0$.

8.2 The remainder in the high indices theorem

The following gap remainder theorem was suggested by Gaier 1966. A proof can be found among other results in Hálasz 1967. Independently I gave a proof by the methods of these notes in a lecture in La Jolla 1966.

THEOREM 8.2 :

Let $(\lambda_n)_0^\infty$ be a sequence of positive real numbers with Hadamard gaps so that $\lambda_{n+1}/\lambda_n \geq c > 1$. If $f(s) = \sum_0^\infty a_n \exp(-\lambda_n s)$ converges for $s > 0$ and $f(s) = O(s^k)$ with a positive k, then $a_n = O(\lambda_n^{-k})$.

We sketch a proof using the basic formula of section 2.2.

With $a(\lambda) = \sum_{\lambda_n \leq \lambda} a_n$ we get $f(s) = \int_0^\infty e^{-s\lambda} da(\lambda)$ and after an integration by parts the standard transformations of our first chapter give

(1) $$\Psi(x) = f(e^{-x}) = K * \phi(x) = O(e^{-kx}) ,$$

where $K(x) = \exp(-x - \exp(-x))$ and $\phi(x) = a(\exp x)$. We know by a lemma of Ingham's (cf. e.g. Hardy 1949) that ϕ is bounded.

The gap condition implies that ϕ is constant in intervals of a length bounded from below.

We apply formula (2.3.1) with $E(y) = \exp(-y^2/\eta)$. Taking Ω as a large but fixed number we prove exactly as in section 2.3 that $Q * \Psi(x) = O(e^{-kx})$. For small y and v we have $\phi(x-v) = \phi(x-y) = \phi(x)$ if x is chosen suitably in the interval of constancy. If Ω is taken sufficiently big, formula (2.3.1) can be written

$$|\phi(x)| \leq r|\phi(x)| + K \sup_{|v| \geq \delta} |\phi(x-v) e^{-v^2/\eta}| + C|Q * \Psi(x)| ,$$

with δ depending on the length of the intervals where ϕ is constant and $r < 1$ so that

(2) $$|\phi(x)| \leq K \sup_{|v|\geq\delta} |\phi(x-v)| \exp(-v^2/\eta) + C\exp(-kx) .$$

If we have taken η so small that

$$K \sup_{|v|\geq\delta} [\exp(kv - v^2/\eta)] = \Theta < 1 ,$$

our theorem follows from (2). We start with any bound A for $\|\phi\|_\infty$ and iterate. The first time (2) reads

$$|\phi(x)| \leq A\Theta + C\exp(-kx) .$$

Introducing this estimate on the right in (2) we get

$$|\phi(x)| \leq A\Theta^2 + CK \sup_{v\geq\delta} |e^{-kx} e^{kv-v^2/\eta}| + C\exp(-kx) = A\Theta + C(1+\Theta)\exp(-kx).$$

Iteration gives

$$|\phi(x)| = A\Theta^{N+1} + C(1 + \Theta + \ldots, + \Theta^N)\exp(-kx)$$

and

$$|\phi(x) \geq C(1 - \Theta)^{-1}\exp(-kx) .$$

The method works in any number of dimensions and the following two--dimensional remainder theorem is proven in Ekhall 1967.

Let the double series $\sum a_{mn} e^{-\lambda_m x - \mu_n y}$ with complex coefficients be absolutely convergent for $(x,y) \in R^2_+$, and suppose that the function f, defined by

$$f(x,y) = \sum_{m=0}^{\infty} \sum_{n=0}^{\infty} a_{mn} e^{-\lambda_m x - \mu_n y}$$

is bounded in R_+^2. Suppose further that the gap conditions

(1) $$\lambda_0 > 0,\ \mu_0 > 0,\ \lambda_{m+1}/\lambda_m \geq c > 1,\ \mu_{n+1}/\mu_n \geq c > 1$$

are satisfied. If $\alpha > 0$, $\beta > 0$ and

(2) $$f(x,y) = O(x^\alpha y^\beta), \quad \text{as} \quad (x,y) \to (0,0)$$

in the region $\{(x,y) \in R_+^2 : x^{1/b_1} > y > x^{b_1}\}$, $b_1 > 1$, then for every b with $1 < b < b_1$, it holds that

$$\sum_{\lambda_m < \lambda} \sum_{\mu_n < \mu} a_{mn} = O(\lambda^{-\alpha}\mu^{-\beta}), \quad \text{as} \quad \lambda, \mu \to \infty$$

in the region $\{(\lambda,\mu)\ R_+^2 : \lambda^{1/b} < \mu < \lambda^b\}$.

8.3 Remainder problems on Z. Renewal theorems

In the introduction to the first chapter we touched on the question of tauberian remainder theorems on other groups. Methods which are very close to those in these notes are applied to a problem on T in Ganelius 1957. Similar results can be obtained on Z and we conclude with one special result of this type, which has a certain resemblance to theorem 2.1.

THEOREM 8.3 :

Let $k \in l^1$ *and* $x \in l^\infty$ *and put* $y = k * x$. *Assume that there is a bounded* u, *holomorphic in a domain* $\{z : 1 < |z| < R\}$, *with boundary values satisfying* $\hat{k}(t)u(e^{it}) = 1$ *for real* t, *where* $\hat{k}(t) = \sum k_m \exp(imt)$.

If $y_k = O(\exp(-W(k))$ *with a positive sub-additive* W *with* $W(k) = 1$ *if* $k \le 0$, *and* $v = \exp(\overline{\lim}\, k^{-1}W(k)) < R$, *then*

$$x_k = O(\exp(-W(k)) .$$

The proof follows since $u(\exp(i.))$ is Fourier transform of an element $u \in l^1$ with $\sum |u_m| \exp(vm) < \infty$. Since $x = u * y$, we get

$$|x_n| = |\sum y_{n-m} u_m| \le \exp(-W(n)) \sum |u_m| \exp W(m) = O(1)\exp(-W(n)) .$$

This theorem can be applied to renewal theorems in the lattice case in a similar way as theorems on R can be applied to other cases. We finally sketch the connection with the following result which is part of a theorem in Stone 1965 (cf. also Essén 1971).

Let F *be a lattice distribution function with lattice constant 1, positive first moment* μ *and finite second moment* σ^2. *If* $1 - F(k) = o(\exp(-rk))$, $k \to \infty$, *for some* $r > 0$, *then the renewal function* G *satisfies*

$$G(k) = (k - 1/2)/\mu + \sigma^2/(2\mu^2) + o(\exp(-sk)) , \quad k \to \infty ,$$

for some positive s.

Let f be the characteristic function of F and define functions H, Q and R by their Fourier-Stieltjes transforms $h(t) = (1-e^{it})^{-1}$, $q(t) = \mu - (1-f(t))/(1-e^{it})$ and $r(t) = q(t)/(1-e^{it})$. Under our assumptions

$$f(t) = 1 - \mu(1-e^{it}) + \tfrac{1}{2}(\sigma^2-\mu)(1-e^{it})^2 + o(t^2) , \quad t \to 0.$$

The identity

$$(1 - f)(g - \mu^{-1}h - \mu^{-2}r) = \sigma^{-2}q^2$$

holds for the Fourier-Stieltjes transform g of G, and hence

$$k * (G - \mu^{-1}H - \mu^{-2}R) = \mu^{-2}Q * Q ,$$

where $\hat{k}(t) = (1 - f(t))/(1 - e^{it})$.

Some computational work gives that the assumption of exponential decrease makes the conditions in theorem 8.3 fulfilled since

$$R(k) = \tfrac{1}{2}(\sigma^2 - \mu) + o(\exp(-sk)) , \quad k \to \infty ,$$
$$Q(k) = o(\exp(-sk)), \; k \to \infty ,$$

for some positive s.

REFERENCES

S. AGMON, T. KANNAI, On the asymptotic behavior of spectral functions and resolvent kernels of elliptic operators, Israel J. Math. 5 (1967), 1 - 30.

V. G. AVAKUMOVIC, Bemerkung über einen Satz des Herrn T. Carleman, Math. Z. 53 (1950), 53 - 58.

V. G. AVAKUMOVIC, Einige Sätze über Laplacesche Integrale, Acad. Serbe Sci. Publ. Inst. Math. 3 (1950), 287 - 304.

G. BERGENDAL, Convergence and summability of eigenfunction expansions connected with elliptic differential operators, Medd. Lunds Univ. Mat. Sem. 14 (1959), 1 - 63.

A. BEURLING, Sur les intégrales de Fourier absolument convergentes, C. R. 9ième congrès des math. scand. Helsinki 1938.

A. BEURLING, Un théorème sur les fonctions bornées et uniformement continues sur l´axe réel, Acta Math. 77 (1945), 127 - 136

G. BERGSTRÖM, Laplace transforms with essential singularities, 'licentiat' thesis, University of Göteborg 1971.

S. BOCHNER, Ein Satz von Landau und Ikehara, Math. Z. 37 (1933), 1 - 9.

S. BOCHNER, K. CHANDRASEKHARAN, Fourier transforms, Princeton 1949.

R. BOJANIC, R. DE VORE, On polynomials of best one sided approximation, L´enseignement math. 12 (1966), 139 - 164.

F. BUREAU, Asymptotic representation of the spectral function of self-adjoint elliptic operators of the second order with variable coefficients I, J. Math. Anal. and appl. 1 (1960), 423 - 483; II, ibid. 4 (1962), 181 - 192.

T. CARLEMAN, L´intégrale de Fourier et questions qui s´y rattachent, Uppsala 1944.

K. CHANDRASEKHARAN, S. MINAKSHISUNDARAM, Typical means, Oxford 1952.

S. A. EKHALL, On the high indices theorem in two dimensions, 'licentiat' thesis, University of Göteborg 1967.

A. ERDELYI (ed.), Tables of integral transforms I, New York 1954.

M. ESSEN, T. GANELIUS, Asymptotic properties of entire functions of order less than one, Am. Math. Soc. Proc. of symposia in pure math. 11 (1968), 193 -201.

M. ESSEN, Banach algebra methods in renewal theory , Stockholm Inst. Techn. 1971.

P. FATOU, Séries trigonométriques et séries de Taylor, Acta Math. 30 (1906), 335-400.

L. FRENNEMO, On general tauberian remainder theorems, Math. Scand. 17 (1965), 77 - 88.

L. FRENNEMO, Tauberian problems for the n-dimensional Laplace transform I, Math. Scand. 19 (1966), 41 - 53; II, ibid. 20 (1967), 225 - 239.

L. FRENNEMO, Studies on the remainder in tauberian problems, Göteborg diss. in science 5 (1967), 12 p.

G. FREUD, Restglied eines Tauberschen Satzes I, Acta Math. Acad. Sci. Hung. 2

(1951), 299 - 308; II, ibid. 3 (1952), 299- 307; III, ibid. 5 (1955), 275 - 288.

G. FREUD, Über einseitige Approximation durch Polynome I, Acta Sci. Math. Szeged 16 (1955), 12 - 28.

G. FREUD, T. GANELIUS, Some remarks on one-sided approximation, Math. Scand. 5 (1957), 276 - 284.

D. GAIER, On the coefficients and the growth of gap power series, J. SIAM Numer. Anal. 3 (1966), 248 - 265.

A. FRIEDMAN, Generalized functions and partial differential equations, Englewood Cliffs 1963.

T. GANELIUS, On the remainder in a tauberian theorem, Kungl. Fysiograf. Sällsk. i Lund Förh. 24 (1954), No. 20, 1 - 6.

T. GANELIUS, On one-sided approximation by trigonometrical polynomials, Math. Scand. 4 (1956), 247 - 258.

T. GANELIUS, Un théorème taubérien pour la transformation de Laplace, C.R. Acad. Sci. Paris 242 (1956), 719 - 721.

T. GANELIUS, Some applications of a lemma on Fourier series, Acad. Serbe. Sci. Publ. Inst. Math. 11 (1957), 9 - 18.

T. GANELIUS, General and special tauberian theorems, C.R. du 13ième congres des math. scand., Helsinki 1958.

T. GANELIUS, The remainder in Wiener's tauberian theorem, Mathematica Gothoburgensia 1 (1962), 13p.

T. GANELIUS, Tauberian theorems for the Stieltjes transform, Math. Scand. 14 (1964), 213 - 219.

T. GANELIUS, An inequality for Stieltjes integrals. To appear in the Proceedings of the 14th Scand. congress of math., Copenhagen 1964.

T. GANELIUS, Some approximation theorems related to Wiener's, to be published in the Proceedings of a conference in Budapest 1969.

I. GELFAND, Normierte Ringe, Matem. Sbornik (51) 9 (1941), 3 - 24.

R. GUSTAFSSON, A note on the behaviour of Fourier transforms at infinity, 'licentiat' thesis, University of Göteborg 1966.

G. HALASZ, Remarks to a paper of D. Gaier on gap theorems, Acta Sci. Math. Szeged 28 (1967), 311 - 322.

G.H. HARDY, A theorem concerning Fourier transforms, J. London Math. Soc. 8 (1933), 227 - 231.

G.H. HARDY, Divergent series, Oxford 1949.

G.H. HARDY, J.E. LITTLEWOOD, Tauberian theorems concerning power series and Dirichlet's series whose terms are positive, Proc. London Math. Soc. 13 (1913), 174 - 191.

I. HIRSCHMAN, On the behaviour of Fourier transforms at infinity and on quasi--analytic classes of functions, Amer. J. Math. 72 (1950), 396 - 406.

I. HIRSCHMAN, D. V. WIDDER, The convolution transform, Princeton 1955.

L. HÖRMANDER, A new proof and a generalization of an inequality of Bohr, Math. Scand. 2 (1954), 33 - 45.

L. HÖRMANDER, On the Riesz means of spectral functions and eigenfuction expansions for elliptic differential operators, Recent advances in the Basic Sciences, Yeshiva University Conference 1966, 155 - 202.

S. IKEHARA, An extension of Landau's theorem in the analytic theorey of numbers, J. Math. and phys. MIT 10 (1931), 1 - 12.

A. INGHAM, Some tauberian theorems connected with the prime number theorem, J. London Math. Soc. 20 (1945), 171 - 180.

A. INGHAM, On tauberian theorems, Proc. London Math. Soc. (3) 14a (1965), 157 - 173.

J. KARAMATA, Über die Hardy-Littlewoodschen Umkehrungen des Abelschen Stetigkeitssatzes, Math. Z. 32 (1930), 319 - 320.

J. KOREVAAR, An estimate of the error in tauberian theorems for power series, Duke Math. J. 18 (1951), 723 - 734.

J. KOREVAAR, Best L_1-approximation and the remainder in Littlewood's theorem, Indag. Math. 15 (1953), 281 - 293.

J. KOREVAAR, A very general form of Littlewood's theorem, ibid. 16 (1954), 36 - 45.

J. KOREVAAR, Another numerical tauberian theorem for power series, ibid. 16 (1954), 46 - 56.

J. KOREVAAR, Distribution proof of Wiener's tauberian theorem, Proc. Am. Math. Soc. 16 (1965), 353 - 355.

N. LEVINSON, Restrictions imposed by certain functions on their Fourier transforms, Duke Math. J. 6 (1940), 722 - 731.

P. LEVY, Sur la convergence absolue des séries de Fourier, Compositio Math. 1 (1934), 1 - 14.

A. LYDEN, A complex tauberian remainder theorem, 'licentiat' thesis, University of Göteborg 1970.

S. LYTTKENS, The remainder in tauberian theorems, Ark. Mat. 2 (1954), 575 - 588.

S. LYTTKENS, The remainder in tauberian theorems II, Ark. Mat. 3 (1956), 315 - 349

P. MALLIAVIN, Un théorème taubérien avec reste pour la transformation de Stieltjes, C. R. Acad. Sci. Paris 255 (1962), 2351 - 2352.

G. W. MORGAN, A note on Fourier transforms, Journal London Math. Soc. 9 (1934), 187 - 192.

R. PALEY, N. WIENER, Fourier transforms in the complex domain, New York 1934.

J. PEETRE, Espaces d'interpolation, généralisations, applications, Rend. Sem. Mat. Fis. Milano 34 (1964), 3 - 34.

H. R. PITT, General tauberian theorems, Proc. London Math. Soc. (2) 44 (1938), 243 - 288.

H. R. PITT, General tauberian theorems II, J. London Math. Soc. 15 (1940), 97 - 112.

H. R. PITT, Tauberian theorems, Oxford 1958.

Å. PLEIJEL, On a theorem by P. Malliavin, Israel J. Math. 1 (1963), 166 - 168.

H. POLLARD, Harmonic analysis of bounded functions, Duke Math. J. 20 (1953), 499 - 512.

W. RUDIN, Real and complex analysis, New York 1966.

R. SCHMIDT, Uber divergente Folgen und lineare Mittelbildungen, Math. Z. 22 (1925), 89 - 152.

L. SCHWARTZ, Théorie des distributions I - II, Paris 1950 - 1951.

V. STENSTRÖM, Sur une certaine classe d'equations intégrales singulières, Math. Scand. 9 (1961), 207 - 228.

C. STONE, On moment generating functions and renewal theory, Ann. Math. Statist. 36 (1965), 1298 - 1301.

M. SUBHANKULOV, Tauberian theorems with remainder term, Amer. Math. Soc. Transl. (2) 26 (1963), 311 - 338.

M. SUBHANKULOV, Some general tauberian theorems with remainder term, Trudy Mat. Inst. Steklov 64 (1961), 239 - 266. (Russian

M. SUBHANKULOV, On a theorem of Littlewood, Izv. Akad. Nauk UzSSR Ser. Fiz.-Mat. Nauk 1964, m. 1, 22 - 30. (Russian)

A. TAUBER, Ein Satz aus der Theorie der unendlichen Reihen, Monatshefte f. Math. 8 (1897), 273 - 277.

E. C. TITCHMARSH, Introduction to the theory of Fourier integrals, 2nd ed., Oxford 1948.

T. VIJAYARAGHAVAN, A tauberian theorem, J. London Math. Soc. 1 (1926), 113 - 120.

H. WIELANDT, Zur Umkehrung des Abelschen Stetigkeitssatzes, Math. Z. 56 (1952), 206 - 207.

N. WIENER, Tauberian theorems, Ann. of Math. (2) 33(1932), 1 - 100.

N. WIENER, The Fourier integral and certain of its applications, Cambridge 1933.

V. VUCKOVIC, Die Stieltjes-Transformation die mit der Geschwindigkeit der Exponentialfunktion unendlich klein wird, Srbska Akad. Nauk. Zbornik Radova 35, Math. Inst. 3 (1953), 255 - 258. (Serbo-croatian)

V. VUCKOVIC, Quelques théorèmes relatifs à la transformation de Stieltjes, Acad. Serbe Sci. Publ. Inst. Math. 6 (1954), 63 - 74.

Lecture Notes in Mathematics

Comprehensive leaflet on request

Vol. 38: R. Berger, R. Kiehl, E. Kunz und H.-J. Nastold, Differentialrechnung in der analytischen Geometrie IV, 134 Seiten. 1967 DM 12,–

Vol. 39: Séminaire de Probabilités I. II, 189 pages. 1967. DM 14,–

Vol. 40: J. Tits, Tabellen zu den einfachen Lie Gruppen und ihren Darstellungen. VI, 53 Seiten. 1967. DM 6.80

Vol. 41: A. Grothendieck, Local Cohomology. VI, 106 pages. 1967. DM 10,–

Vol. 42: J. F. Berglund and K. H. Hofmann, Compact Semitopological Semigroups and Weakly Almost Periodic Functions. VI, 160 pages. 1967. DM 12,–

Vol. 43: D. G. Quillen, Homotopical Algebra. VI, 157 pages. 1967. DM 14,–

Vol. 44: K. Urbanik, Lectures on Prediction Theory. IV, 50 pages. 1967. DM 5,80

Vol. 45: A. Wilansky, Topics in Functional Analysis. VI, 102 pages. 1967. DM 9,60

Vol. 46: P. E. Conner, Seminar on Periodic Maps. IV, 116 pages. 1967. DM 10,60

Vol. 47: Reports of the Midwest Category Seminar I. IV, 181 pages. 1967. DM 14,80

Vol. 48: G. de Rham, S. Maumary et M. A. Kervaire, Torsion et Type Simple d'Homotopie. IV, 101 pages. 1967. DM 9,60

Vol. 49: C. Faith, Lectures on Injective Modules and Quotient Rings. XVI, 140 pages. 1967. DM 12,80

Vol. 50: L. Zalcman, Analytic Capacity and Rational Approximation. VI, 155 pages. 1968. DM 13.20

Vol. 51: Séminaire de Probabilités II. IV, 199 pages. 1968. DM 14,–

Vol. 52: D. J. Simms, Lie Groups and Quantum Mechanics. IV, 90 pages. 1968. DM 8,–

Vol. 53: J. Cerf, Sur les difféomorphismes de la sphère de dimension trois ($\Gamma_4 = 0$). XII, 133 pages. 1968. DM 12,–

Vol. 54: G. Shimura, Automorphic Functions and Number Theory. VI, 69 pages. 1968. DM 8,–

Vol. 55: D. Gromoll, W. Klingenberg und W. Meyer, Riemannsche Geometrie im Großen. VI, 287 Seiten. 1968. DM 20,–

Vol. 56: K. Floret und J. Wloka, Einführung in die Theorie der lokalkonvexen Räume. VIII, 194 Seiten. 1968. DM 16,–

Vol. 57: F. Hirzebruch und K. H. Mayer, O (n)-Mannigfaltigkeiten, exotische Sphären und Singularitäten. IV, 132 Seiten. 1968. DM 10,80

Vol. 58: Kuramochi Boundaries of Riemann Surfaces. IV, 102 pages. 1968. DM 9,60

Vol. 59: K. Jänich, Differenzierbare G-Mannigfaltigkeiten. VI, 89 Seiten. 1968. DM 8,–

Vol. 60: Seminar on Differential Equations and Dynamical Systems. Edited by G. S. Jones. VI, 106 pages. 1968. DM 9,60

Vol. 61: Reports of the Midwest Category Seminar II. IV, 91 pages. 1968. DM 9,60

Vol. 62: Harish-Chandra, Automorphic Forms on Semisimple Lie Groups X, 138 pages. 1968. DM 14,–

Vol. 63: F. Albrecht, Topics in Control Theory. IV, 65 pages. 1968. DM 6,80

Vol. 64: H. Berens, Interpolationsmethoden zur Behandlung von Approximationsprozessen auf Banachräumen. VI, 90 Seiten. 1968. DM 8,–

Vol. 65: D. Kölzow, Differentiation von Maßen. XII, 102 Seiten. 1968. DM 8,–

Vol. 66: D. Ferus, Totale Absolutkrümmung in Differentialgeometrie und -topologie. VI, 85 Seiten. 1968. DM 8,–

Vol. 67: F. Kamber and P. Tondeur, Flat Manifolds. IV, 53 pages. 1968. DM 5,80

Vol. 68: N. Boboc et P. Mustată, Espaces harmoniques associés aux opérateurs différentiels linéaires du second ordre de type elliptique. VI, 95 pages. 1968. DM 8,60

Vol. 69: Seminar über Potentialtheorie. Herausgegeben von H. Bauer. VI, 180 Seiten. 1968. DM 14,80

Vol. 70: Proceedings of the Summer School in Logic. Edited by M. H. Löb. IV, 331 pages. 1968. DM 20,–

Vol. 71: Séminaire Pierre Lelong (Analyse), Année 1967 – 1968. VI, 190 pages. 1968. DM 14,

Vol. 72: The Syntax and Semantics of Infinitary Languages. Edited by J. Barwise. IV, 268 pages. 1968. DM 18,–

Vol. 73: P. E. Conner, Lectures on the Action of a Finite Group. IV, 123 pages. 1968. DM 10,–

Vol. 74: A. Fröhlich, Formal Groups. IV, 140 pages. 1968. DM 12,–

Vol. 75: G. Lumer, Algèbres de fonctions et espaces de Hardy. VI, 80 pages. 1968. DM 8,–

Vol. 76: R. G. Swan, Algebraic K-Theory. IV, 262 pages. 1968. DM 18,–

Vol. 77: P.-A. Meyer, Processus de Markov: la frontière de Martin. IV, 123 pages. 1968. DM 10,–

Vol. 78: H. Herrlich, Topologische Reflexionen und Coreflexionen. XVI, 166 Seiten. 1968. DM 12,–

Vol. 79: A. Grothendieck, Catégories Cofibrées Additives et Complexe Cotangent Relatif. IV, 167 pages. 1968. DM 12,–

Vol. 80: Seminar on Triples and Categorical Homology Theory. Edited by B. Eckmann. IV, 398 pages. 1969. DM 20,–

Vol. 81: J.-P. Eckmann et M. Guenin, Méthodes Algébriques en Mécanique Statistique. VI, 131 pages. 1969. DM 12,–

Vol. 82: J. Wloka, Grundräume und verallgemeinerte Funktionen. VIII, 131 Seiten. 1969. DM 12,–

Vol. 83: O. Zariski, An Introduction to the Theory of Algebraic Surfaces. IV, 100 pages. 1969. DM 8,–

Vol. 84: H. Lüneburg, Transitive Erweiterungen endlicher Permutationsgruppen. IV, 119 Seiten. 1969. DM 10.–

Vol. 85: P. Cartier et D. Foata, Problèmes combinatoires de commutation et réarrangements. IV, 88 pages. 1969. DM 8,–

Vol. 86: Category Theory, Homology Theory and their Applications I. Edited by P. Hilton. VI, 216 pages. 1969. DM 16,–

Vol. 87: M. Tierney, Categorical Constructions in Stable Homotopy Theory. IV, 65 pages. 1969. DM 6,–

Vol. 88: Séminaire de Probabilités III. IV, 229 pages. 1969. DM 18,–

Vol. 89: Probability and Information Theory. Edited by M. Behara, K. Krickeberg and J. Wolfowitz. IV, 256 pages. 1969. DM 18,–

Vol. 90: N. P. Bhatia and O. Hajek, Local Semi-Dynamical Systems. II, 157 pages. 1969. DM 14,–

Vol. 91: N. N. Janenko, Die Zwischenschrittmethode zur Lösung mehrdimensionaler Probleme der mathematischen Physik. VIII, 194 Seiten. 1969. DM 16,80

Vol. 92: Category Theory, Homology Theory and their Applications II. Edited by P. Hilton. V, 308 pages. 1969. DM 20,–

Vol. 93: K. R. Parthasarathy, Multipliers on Locally Compact Groups. III, 54 pages. 1969. DM 5,60

Vol. 94: M. Machover and J. Hirschfeld, Lectures on Non-Standard Analysis. VI, 79 pages. 1969. DM 6,–

Vol. 95: A. S. Troelstra, Principles of Intuitionism. II, 111 pages. 1969. DM 10,–

Vol. 96: H.-B. Brinkmann und D. Puppe, Abelsche und exakte Kategorien, Korrespondenzen. V, 141 Seiten. 1969. DM 10,–

Vol. 97: S. O. Chase and M. E. Sweedler, Hopf Algebras and Galois theory. II, 133 pages. 1969. DM 10,–

Vol. 98: M. Heins, Hardy Classes on Riemann Surfaces. III, 106 pages. 1969. DM 10,–

Vol. 99: Category Theory, Homology Theory and their Applications III. Edited by P. Hilton. IV, 489 pages. 1969. DM 24,–

Vol. 100: M. Artin and B. Mazur, Etale Homotopy. II, 196 Seiten. 1969. DM 12,–

Vol. 101: G. P. Szegö et G. Treccani, Semigruppi di Trasformazioni Multivoche. VI, 177 pages. 1969. DM 14,–

Vol. 102: F. Stummel, Rand- und Eigenwertaufgaben in Sobolewschen Räumen. VIII, 386 Seiten. 1969. DM 20,–

Vol. 103: Lectures in Modern Analysis and Applications I. Edited by C. T. Taam. VII, 162 pages. 1969. DM 12,–

Vol. 104: G. H. Pimbley, Jr., Eigenfunction Branches of Nonlinear Operators and their Bifurcations. II, 128 pages. 1969. DM 10,–

Vol. 105: R. Larsen, The Multiplier Problem. VII, 284 pages. 1969. DM 18,–

Vol. 106: Reports of the Midwest Category Seminar III. Edited by S. Mac Lane. III, 247 pages. 1969. DM 16,–

Vol. 107: A. Peyerimhoff, Lectures on Summability. III, 111 pages. 1969. DM 8,–

Vol. 108: Algebraic K-Theory and its Geometric Applications. Edited by R. M.F. Moss and C. B. Thomas. IV, 86 pages. 1969. DM 6,–

Vol. 109: Conference on the Numerical Solution of Differential Equations. Edited by J. Ll. Morris. VI, 275 pages. 1969. DM 18,–

Vol. 110: The Many Facets of Graph Theory. Edited by G. Chartrand and S. F. Kapoor. VIII, 290 pages. 1969. DM 18,–

Vol. 111: K. H. Mayer, Relationen zwischen charakteristischen Zahlen. III, 99 Seiten. 1969. DM 8,–

Vol. 112: Colloquium on Methods of Optimization. Edited by N. N. Moiseev. IV, 293 pages. 1970. DM 18,–

Vol. 113: R. Wille, Kongruenzklassengeometrien. III, 99 Seiten. 1970. DM 8,–

Vol. 114: H. Jacquet and R. P. Langlands, Automorphic Forms on GL (2). VII, 548 pages. 1970. DM 24,–

Vol. 115: K. H. Roggenkamp and V. Huber-Dyson, Lattices over Orders I. XIX, 290 pages. 1970. DM 18,–

Vol. 116: Séminaire Pierre Lelong (Analyse) Année 1969. IV, 195 pages. 1970. DM 14,–

Vol. 117: Y. Meyer, Nombres de Pisot, Nombres de Salem et Analyse Harmonique. 63 pages. 1970. DM 6.–

Vol. 118: Proceedings of the 15th Scandinavian Congress, Oslo 1968. Edited by K. E. Aubert and W. Ljunggren. IV, 162 pages. 1970. DM 12,–

Vol. 119: M. Raynaud, Faisceaux amples sur les schémas en groupes et les espaces homogènes. III, 219 pages. 1970. DM 14,–

Vol. 120: D. Siefkes, Büchi's Monadic Second Order Successor Arithmetic. XII, 130 Seiten. 1970. DM 12,–

Vol. 121: H. S. Bear, Lectures on Gleason Parts. III, 47 pages. 1970. DM 6,–

Vol. 122: H. Zieschang, E. Vogt und H.-D. Coldewey, Flächen und ebene diskontinuierliche Gruppen. VIII, 203 Seiten. 1970. DM 16,–

Vol. 123: A. V. Jategaonkar, Left Principal Ideal Rings. VI, 145 pages. 1970. DM 12,–

Vol. 124: Séminare de Probabilités IV. Edited by P. A. Meyer. IV, 282 pages. 1970. DM 20,–

Vol. 125: Symposium on Automatic Demonstration. V, 310 pages. 1970. DM 20,–

Vol. 126: P. Schapira, Théorie des Hyperfonctions. XI, 157 pages. 1970. DM 14,–

Vol. 127: I. Stewart, Lie Algebras. IV, 97 pages. 1970. DM 10,–

Vol. 128: M. Takesaki, Tomita's Theory of Modular Hilbert Algebras and its Applications. II, 123 pages. 1970. DM 10,–

Vol. 129: K. H. Hofmann, The Duality of Compact Semigroups and C*- Bigebras. XII, 142 pages. 1970. DM 14,–

Vol. 130: F. Lorenz, Quadratische Formen über Körpern. II, 77 Seiten. 1970. DM 8,–

Vol. 131: A Borel et al., Seminar on Algebraic Groups and Related Finite Groups. VII, 321 pages. 1970. DM 22,–

Vol. 132: Symposium on Optimization. III, 348 pages. 1970. DM 22,–

Vol. 133: F. Topsøe, Topology and Measure. XIV, 79 pages. 1970. DM 8,–

Vol. 134: L. Smith, Lectures on the Eilenberg-Moore Spectral Sequence. VII, 142 pages. 1970. DM 14,–

Vol. 135: W. Stoll, Value Distribution of Holomorphic Maps into Compact Complex Manifolds. II, 267 pages. 1970. DM 18,–

Vol. 136: M. Karoubi et al., Séminaire Heidelberg-Saarbrücken-Strasbuorg sur la K-Théorie. IV, 264 pages. 1970. DM 18,–

Vol. 137: Reports of the Midwest Category Seminar IV. Edited by S. MacLane. III, 139 pages. 1970. DM 12,–

Vol. 138: D. Foata et M. Schützenberger, Théorie Géométrique des Polynômes Eulériens. V, 94 pages. 1970. DM 10,–

Vol. 139: A. Badrikian, Séminaire sur les Fonctions Aléatoires Linéaires et les Mesures Cylindriques. VII, 221 pages. 1970. DM 18,–

Vol. 140: Lectures in Modern Analysis and Applications II. Edited by C. T. Taam. VI, 119 pages. 1970. DM 10,–

Vol. 141: G. Jameson, Ordered Linear Spaces. XV, 194 pages. 1970. DM 16,–

Vol. 142: K. W. Roggenkamp, Lattices over Orders II. V, 388 pages. 1970. DM 22,–

Vol. 143: K. W. Gruenberg, Cohomological Topics in Group Theory. XIV, 275 pages. 1970. DM 20,–

Vol. 144: Seminar on Differential Equations and Dynamical Systems, II. Edited by J. A. Yorke. VIII, 268 pages. 1970. DM 20,–

Vol. 145: E. J. Dubuc, Kan Extensions in Enriched Category Theory. XVI, 173 pages. 1970. DM 16,–

Vol. 146: A. B. Altman and S. Kleiman, Introduction to Grothendieck Duality Theory. II, 192 pages. 1970. DM 18,–

Vol. 147: D. E. Dobbs, Cech Cohomological Dimensions for Commutative Rings. VI, 176 pages. 1970. DM 16,–

Vol. 148: R. Azencott, Espaces de Poisson des Groupes Localement Compacts. IX, 141 pages. 1970. DM 14,–

Vol. 149: R. G. Swan and E. G. Evans, K-Theory of Finite Groups and Orders. IV, 237 pages. 1970. DM 20,–

Vol. 150: Heyer, Dualität lokalkompakter Gruppen. XIII, 372 Seiten. 1970. DM 20,–

Vol. 151: M. Demazure et A. Grothendieck, Schémas en Groupes I. (SGA 3). XV, 562 pages. 1970. DM 24,–

Vol. 152: M. Demazure et A. Grothendieck, Schémas en Groupes II. (SGA 3). IX, 654 pages. 1970. DM 24,–

Vol. 153: M. Demazure et A. Grothendieck, Schémas en Groupes III. (SGA 3). VIII, 529 pages. 1970. DM 24,–

Vol. 154: A. Lascoux et M. Berger, Variétés Kähleriennes Compactes. VII, 83 pages. 1970. DM 8,–

Vol. 155: Several Complex Variables I, Maryland 1970. Edited by J. Horváth. IV, 214 pages. 1970. DM 18,–

Vol. 156: R. Hartshorne, Ample Subvarieties of Algebraic Varieties. XIV, 256 pages. 1970. DM 20,–

Vol. 157: T. tom Dieck, K. H. Kamps und D. Puppe, Homotopietheorie. VI, 265 Seiten. 1970. DM 20,–

Vol. 158: T. G. Ostrom, Finite Translation Planes. IV. 112 pages. 1970. DM 10,–

Vol. 159: R. Ansorge und R. Hass. Konvergenz von Differenzenverfahren für lineare und nichtlineare Anfangswertaufgaben. VIII, 145 Seiten. 1970. DM 14,–

Vol. 160: L. Sucheston, Constributions to Ergodic Theory and Probability. VII, 277 pages. 1970. DM 20,–

Vol. 161: J. Stasheff, H-Spaces from a Homotopy Point of View. VI, 95 pages. 1970. DM 10,–

Vol. 162: Harish-Chandra and van Dijk, Harmonic Analysis on Reductive p-adic Groups. IV, 125 pages. 1970. DM 12,–

Vol. 163: P. Deligne, Equations Différentielles à Points Singuliers Reguliers. III, 133 pages. 1970. DM 12,–

Vol. 164: J. P. Ferrier, Seminaire sur les Algebres Complètes. II, 69 pages. 1970. DM 8,–

Vol. 165: J. M. Cohen, Stable Homotopy. V, 194 pages. 1970. DM 16,–

Vol. 166: A. J. Silberger, PGL_2 over the p-adics: its Representations, Spherical Functions, and Fourier Analysis. VII, 202 pages. 1970. DM 18,–

Vol. 167: Lavrentiev, Romanov and Vasiliev, Multidimensional Inverse Problems for Differential Equations. V, 59 pages. 1970. DM 10,–

Vol. 168: F. P. Peterson, The Steenrod Algebra and its Applications: A conference to Celebrate N. E. Steenrod's Sixtieth Birthday. VII, 317 pages. 1970. DM 22,–

Vol. 169: M. Raynaud, Anneaux Locaux Henséliens. V, 129 pages. 1970. DM 12,–

Vol. 170: Lectures in Modern Analysis and Applications III. Edited by C. T. Taam. VI, 213 pages. 1970. DM 18,–

Vol. 171: Set-Valued Mappings, Selections and Topological Properties of 2^X. Edited by W. M. Fleischman. X, 110 pages. 1970. DM 12,–

Vol. 172: Y.-T. Siu and G. Trautmann, Gap-Sheaves and Extension of Coherent Analytic Subsheaves. V, 172 pages. 1971. DM 16,–

Vol. 173: J. N. Mordeson and B. Vinograde, Structure of Arbitrary Purely Inseparable Extension Fields. IV, 138 pages. 1970. DM 14,–

Vol. 174: B. Iversen, Linear Determinants with Applications to the Picard Scheme of a Family of Algebraic Curves. VI, 69 pages. 1970. DM 8,–

Vol. 175: M. Brelot, On Topologies and Boundaries in Potential Theory. VI, 176 pages. 1971. DM 18,–

Vol. 176: H. Popp, Fundamentalgruppen algebraischer Mannigfaltigkeiten. IV, 154 Seiten. 1970. DM 16,–

Vol. 177: J. Lambek, Torsion Theories, Additive Semantics and Rings of Quotients. VI, 94 pages. 1971. DM 12,–

Vol. 178: Th. Bröcker und T. tom Dieck, Kobordismentheorie. XVI, 191 Seiten. 1970. DM 18,–

Vol. 179: Seminaire Bourbaki – vol. 1968/69. Exposés 347-363. IV. 295 pages. 1971. DM 22,–

Vol. 180: Séminaire Bourbaki – vol. 1969/70. Exposés 364-381. IV, 310 pages. 1971. DM 22,–

Vol. 181: F. DeMeyer and E. Ingraham, Separable Algebras over Commutative Rings. V, 157 pages. 1971. DM 16.–